EXERCISES IN GIS
to accompany

FUNDAMENTALS OF GEOGRAPHIC INFORMATION SYSTEMS

SECOND EDITION

Michael N. DeMers
New Mexico State University

JOHN WILEY & SONS, INC.

NEW YORK • CHICHESTER • WEINHEIM • BRISBANE • SINGAPORE • TORONTO

To order books or for customer service call 1-800-CALL-WILEY (225-5945).

ISBN 0-471-35267-5

Printed in the United States of America

10 9 8 7 6 5 4 3 2 1

Printed and bound by Victor Graphics, Inc.

Acknowledgements

I would like to thank all of the teaching assistants I have had over the past three years here at New Mexico State University for providing important corrections and insights for these laboratory exercises. They include James Hangen, Sonya Krogh, and Jennifer Scheffel. Mr. Hangen also provided the Gove County database based on his laboratory experiences. Barbara Nolen, the GIS coordinator for the Jornada LTER at New Mexico State University, also deserves thanks for important suggestions and error corrections. I very much appreciate the editorial work of Lin Liu at the University of Cincinnati, Steven R. Hoffer at the Community College of Southern Nevada, and Douglas J. Wheeler of George Mason University for the detailed, invaluable comments they provided during the production of this manual. Without their help this manual would not exist. Finally I want to thank the students who have been subjected to the early versions of this manual. Their patience, and useful comments are especially appreciated because I learned so much about teaching the laboratory from them.

Contents

Laboratory Exercise 1

GIS Resources and Applications

Introduction

GIS is growing all the time as a tool with many applications, many opportunities for employment, a wide variety of available software, new vendors, and a vast and growing array of demonstrations, teaching materials, books, bibliographies and many other resources. Because these are all growing at an alarming rate it is impossible to keep in touch with all of them at a single location. Instead, we need to periodically keep up with what is going on through the electronic media (the web) because many of these changes will be reflected in the changing resources found there. When you are finished with this laboratory exercise you should have begun to see what some of these resources are and will be able to update your own knowledge base through continued use of these resources.

Learning Objectives:

This first exercise is designed to get you thinking about what resources are available to you (e.g. data, demos, teaching, books, bibliographies, etc), who is using GIS (vendors, applications companies, government agencies, etc.), what jobs are available for GIS professionals, and what skills they might be looking for. To do this you are going to use currently available Internet technology (the Web). You are going to search the web for information, thus traveling to different sources of information right from your desktop computer. This is not the only method of acquiring this information, as you will see, but it is certainly the easiest, as long as you have the technology.

When you are finished with this exercise you should be able to:

1. Locate sources of digital data for GIS.
2. Identify, download, and display GIS demonstrations.
3. Download teaching (and learning) aids, bibliographies and other resources.
4. Identify specific GIS vendors, application companies, agencies, and others using GIS.
5. List GIS jobs, position titles, salaries, and skills needed for each position.

Linkage to your text:

Chapter 1 of Fundamentals of Geographic Information Systems lists 6 major learning objectives. Among these are defining what a GIS is (#1), describing the capabilities of modern GIS (#5), and enumerating potential users of GIS (#6). While all 6 of these objectives can largely be gained by reading the text, the latter three are rather more difficult to get at without some external sources to peruse. This exercise is meant to enhance your ability to gain more insights into these concepts in addition to what is available through your text.

Methods:

We begin this exercise by first visiting the GIS resources page available at New Mexico State University's (NMSU) geography department. When you open your web browser go to the NMSU web page found at www.nmsu.edu. Navigate to the **Academics** button. From there you will navigate through **Arts and Sciences**, then to **Academic Departments**, and finally to **Geography**. When you get there be sure to click on the Department rather than on the department chair's e-mail address. Once inside the geography department webpage, go to the GIS resources button on the left. There are many links to some wonderful GIS sites there. They are organized by topic for your convenience.

After you have visited this site you might try going back to the beginning and using one of the search engines (e.g. Yahoo, Lycos, etc.) available through the search button on NMSU's webpage or your own webpages if you wish. There are also some specific web sites you should check out. For example, the following two URL's have specific information about the history of GIS.

http://www.geog.buffalo.edu/ncgia/gishist

http://www.casa.ucl.ac.uk/gistimeline/gistimeline.html

The first of these two provides a rather nice discussion of the history of GIS in more detail that your text. On the second webpage you will find a compact approach to the history of GIS in the form of a GIS timeline.

Also, the USGS has a basic introduction to GIS webpage at...

http://info.er.usgs.gov/research/gis/title.html

You might also want to link to ESRI's fullsome webpage at...

http://www.esri.com

and the Association of American Geographers' GIS specialty group webpage as well...

http://www.cla.sc.edu/gis/aaggis.html

As you visit each link, make bookmarks for sites that you find interesting and for which you might be able to gain insights, data, jobs, software, hardware, etc. Keep your links on a floppy disk so that you will be able to use them yourself and provide copies to you laboratory instructor. (Your laboratory instructor may want to compile the best of the best and update the GIS links in your department to provide a better service to you and your classmates. Or you may wish to create your own list of GIS links for future reference).

Products:

The following should be turned in to your laboratory instructor as final products for the laboratory exercise:

1. A 3.5" DSHD floppy with the bookmarks for the most interesting sites you found on the web.

2. A printed document displaying all the bookmarks you found, placed in the following categories.

4

- Dead links (links that are no longer functioning)

- Good sources of data (where you can download, trade, or buy spatial data)

- Vendors (where you can learn about the GIS software available to you)

- Free software (where you can download free GIS software packages, demos, or prototypes)

- Applications companies

- Job sources

3. Depending on your instructor, the laboratories may or may not be graded. Your laboratory instructor may want to keep track of all laboratories completed and will be encouraged to include this information as part of your laboratory record or laboratory grade.

Laboratory Exercise 2

Getting Started With ArcView®

Introduction:

ArcView 3 is a professional level GIS from Environmental Systems Research Institute (ESRI) in Redlands, California. While it may lack some of the functionality of its larger cousin (ARC/INFO), it maintains a subtantial amount of that functionality without ARC/INFO 7.2's difficult user interface and steep learning curve. The advantages of ArcView's windows based user interface are, however, often counterbalanced by the ease with which a user can get lost amid the icons, pulldown menus and default subdirectory system. This second exercise addresses the user interface and functionality issues of the software. In laboratory 3 we will look further at the directory structure.

Learning Objectives:

The purpose of this laboratory exercise is to allow you a gentle introduction to what the software is capable of, what the icons mean and what they do. In addition to learning some of the basics of ArcView 3, you will also examine the concept of the map in a digital framework. You will look at the concepts of scale and projection, map symbology, generalization, legends, and the concept of the analytical (holistic) paradigm.

When you are finished with this exercise you should be able to:

1. Navigate in and out of ArcView and recognize and use some of the basic icons.

2. Identify the scale of data measurement for a variety of point, line, and polygon features.

3. Describe the difference between the communication paradigm and the analytical or holistic paradigm.

4. Describe some of the different map projections used and enumerate the properties preserved by each.

5. Identify some of the grid systems used and link them to their map projections.

6. List, define, and explain some important spatial vocabulary used with GIS software.

Linkage to your text:

Most of the learning objectives from Chapter 2 and some from Chapter 3 in Fundamentals of Geographic Information Systems are examined in this exercise. For example, you will improve your geographic vocabulary (geographic filter) (#1) through the use of geographic terms inside ArcView, by examining GIS *coverages* or *themes* (individual thematic maps) at different scales you will see the impact of scale on the perception of our world (#2), by looking at both the map features (points, lines and polygons as in #3) and their descriptors you will concentrate on spatial data rather than aspatial data. As you become more involved in model building you will have a chance to revisit the idea of aspatial data. Also in this exercise you will see examples of point, line and area features in a number of measurement levels (#4, and #5), and you will examine grid systems and projections (Chapter 3). You will also look at the analytical paradigm as opposed to the traditional (communication) paradigm (Chapter 3).

Methods:

Turn on your computer and start ArcView. To start ArcView in the Windows environment there are three approaches. The first is to go to the start button on the left of the tool bar and click (and hold) then go through the side bar menus until you find ArcView 3, when you find it, click on it and the software will begin. A second method is to go to the icon that says "my computer" and click on that. By following the appropriate folders and subdirectories you will find and locate the ArcView 3 icon and click it to start. A third method would be to create a shortcut to ArcView 3. Your laboratory instructor will determine whether or not he/she wants you to use this approach and will, if needed, explain how this is done.

- Your instructors may have set up a shortcut to ArcView so you may not need to use the methods mentioned above. Just use the shortcut. It should look something like this:

When ArcView loads up you will see a screen that looks something like this:

You notice that there is a welcome window in the center of the ArcView window that shows up the first time you use ArcView. You see at the bottom there is a check box that can eliminate this welcome window from showing up. There are also three additional check boxes that give you the options of creating a new view (window), a blank project, or an existing project. For the time being we'll check the second box for a blank project. This removes the welcome window.

Notice that there are icons on the top and, within a smaller window, vertically along the left side. Also note that the right side of the screen is blank. This is where your project maps will be displayed. A **project** is an empty space where you will save all your work during a **session** . You begin an ArcView session by first opening a project (either existing or new) from the pulldown menus at the top where it says "File."

Go to "file" and the pull down menu will give you options for creating a new project or opening an old project. Choose the "open project" and select "avtutor" and then scroll down and select "arcview." Next a file named "qstart" appears in the left window. Click on it to start your session.

When you do this the three selectable views that will open are "Atlanta," "United States," and "World." Double click on World and see what happens. A new window (**view**) opens up to show you a map of the world. Now it's time to wander around in the software to see what can be seen.

Here are some tasks that you should employ to get familiar with the software and some questions that require you to learn some of what the software can do.

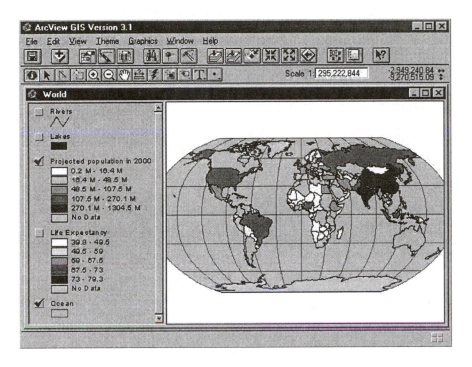

- First, let's see what our screen has to say about the scale of your map. Without making any of your windows larger write down what the scale of the map is here _____ using the representative fraction method. You should find this information in the upper right hand corner of the view frame.

- There are a number of ways to zoom in (change to a larger scale). Try some out. For example, you can use the zoom icons that include a magnifying glass with a plus (+) sign for  zooming in and one with a minus (-) sign for zooming out. Try zooming in by selecting the + magnifyer and going to the center of your world map and clicking once. What is the new scale (representative fraction)? _____ Let's say you want to see the map at a specific scale (say 1:24,000) you can do this by highlighting the denominator and typing in 24,000. Try it out to see what it looks like. There is also a set of icons in the upper right center of the window that allow you to zoom in and out as well. Try them out to see what they do (but first, place the cursor over each one and wait a second or two to see what the caption says the icon does).

- Now that you have located these icons, you will notice that there are a couple that are not active. The reason for this is that they both require that a certain *theme* (*coverage*) be active before they work. A theme or coverage as we will call it in this course is a single thematic map that displays individual topics (themes) related to the overall study area. In ArcView themes can only be created using a particular data entity type (i.e. only for points or only for lines, or only for polygons). In addition, to make things simple, the themes are usually very specific in their topic (e.g. the points will usually all be the same type of points, such as the locations of wells, rather than wells, homes, trees, etc.) When there is an active theme the remaining zoom icons will become active. To activate a coverage theme such as "projected population 2000" go to the legend at left and click the left mouse button on that theme. Note how the theme appears to have become raised relative to the other themes. You should also note that the theme has a small check box within it. This is a toggle switch that turns the selected features on or off for display. Try selecting and deselecting the coverages and their display to see what you get. Now go back and test out the zoom methods using those icons that were previously inactive. When you are done make some notes about what each icon does.

- You should remember that we are looking at at particular view "world" within the project. It would be nice if we could find out more about what this view is like. Go to the "view" pulldown menu and select "properties" from it. What information is there? What is missing? What is the

map projection? What information is in the "area of interest" pulldown menu? What you are looking at is a *data dictionary* , a concept that we will return to at a later time. Click OK.

- Try zooming into Europe a few times. Notice how the seemingly smooth lines are becoming more jagged. Can you explain this? A hint is that it has to do with input into the GIS. We will look more closely at this in the future.

- If you examined the "view" properties, you should have seen that the map was presented in the Robinson projection. One of the nice features of a GIS is that you can change projections, although it will introduce some error into the resulting map. This error is probably not as important as some that will creep into the maps through the input process and the map production process that preceded it. Click on the projection icon inside the properties menu. You will notice a number of alternative map projections to choose from. Try some of these out and note the changes that take place in the appearance of the map and the relationships between the meridians and parallels. These changes help define what type of projection you have and the properties that are affected by the projection itself. Refer to your text and see what insights are there. You might also want to surf the great globes webpage for some rather nice examples of the major types of map projections. The URL is http://www.hum.amu.edu.pl/~zbzw/glob/globl.htm. Explanations for them can be found in a number of cartography texts as well as the USGS document listed at the bottom of the map projections on the webpage.

- There are a number of other things that you need to examine in this second laboratory. Pointing, but not clicking, go through both the upper and the lower bank of icons and read what the caption says about each icon. Like most windows-based software ArcView has this handy feature, but it is cryptic, and not always as useful as it should be. Once you have done this activate the lakes theme by checking near the term lakes (check the box as well), and zoom in on the United States. Click on the identify button, then click on any of the Great Lakes. Notice what this does, it shows us the **attribute** data attached to the **entity** data (in this case polygonal data).

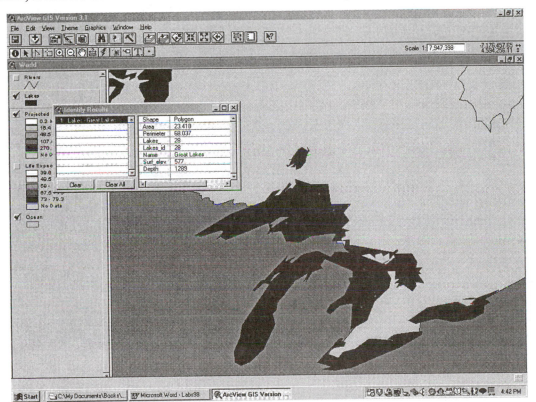

- There is another way to get acquainted with the attribute data attached to the entity data as well, and this will allow you to view a large number of attributes at a time. But, before we do that we have to select a theme to view and check mark it for display as above. Take the "world" view and select the lakes as the active theme. Make sure you check mark it for display. You should see a number of lakes showing up on the screen as blue polygons. Now go to the set of icons on the top of your screen and click on the tables icon. You should see a large array of tabular data for all the lakes included in the database. Peruse this list and identify what attributes are available for this theme of the database. In a later exercise we will show you how to select individual lakes, rather than all of them, using both graphic and tabular search methods and we will show you how to "promote" the ones you selected so you don't have to search through all of them for the ones you really want to look at. When you are done perusing this set of data go to "file" and click on close (not quit). This will return you to your original world view.

ArcView GIS Version 3.1

File Edit Table Field Window Help

0 of   33 selected

Attributes of Lakes

Shape	Area	Perimeter	Lakes	Lakes_id	Name	Surf_area	Depth
Polygon	1.715	15.600	1	1	Lake Onega	108	394
Polygon	1.056	9.754	2	2	Reindeer Lake	1106	720
Polygon	4.361	18.503	3	3	Lake Baikal	1493	5318
Polygon	2.092	16.853	4	4	Lake Balkhash	1115	87
Polygon	0.640	4.904	5	5	Issyk Kul	5279	2303
Polygon	1.917	7.678	6	6	Lake Chad	787	24
Polygon	0.253	1.993	7	7	Lake Tana	6003	30
Polygon	0.600	5.210	8	8	Lake Turkana	1230	720
Polygon	5.689	17.807	9	9	Lake Victoria	3720	277
Polygon	2.671	13.778	10	10	Lake Tanganyika	2543	4800
Polygon	3.001	9.047	11	11	Lake Ladoga	13	755
Polygon	0.916	5.805	12	12	Lake Vanern	144	325
Polygon	0.638	4.514	13	13	Great Salt Lake	4200	48
Polygon	0.459	2.998	14	14	Koko Nor	10515	125
Polygon	0.435	3.602	15	15	Lake Albert	2030	168
Polygon	2.135	11.608	16	16	Lake Nyasa	1550	2280
Polygon	0.606	6.192	17	17	Lake Titicaca	12500	990
Polygon	0.396	3.695	18	18	Lake Urmia	4180	49
Polygon	7.737	19.345	19	19	Aral Sea	174	220

- In the world view all of the coverages were area or polygon (2-D) objects. Let's look at a view where we have some line objects to look at. First, "close" the world view. This will return you to the screen shown earlier in your laboratory exercise (page 9). Double click on the "United States" view and see what comes up there. As before, check the scale, examine the contents of the view data dictionary, see what information is contained there, etc. Make the "highways" theme active and viewable by clicking on that portion of the legend and clicking on the small square. You now have a set of linear (2-D) objects which, when connected, make up a **network**. Use the table icon to identify the contents of this theme as you did with the lakes theme in the world view.

- Repeat the last set of instructions for the Atlanta database to see what is available there. As before check out the "view" properties to see what is available for this view. Activate individual themes and check out the "theme" properties as well. You should note that there are no point objects in the databases you have examined. As you will notice from your text, point objects are not as commonly employed in GIS as are polygon and line objects. We'll see some later on in the laboratory exercises.

Before you quit ArcView look at some of the review questions on the next page.

Review and Discussion Questions.

Now that you have viewed all three of the databases you might want to revisit them to answer some of the questions below. Some of the questions are designed to get you thinking rather than to elicit specific answers so take some time to relate them to your reading and the lecture material.

- What level of data measurement did you find for lakes, roads, etc.? Take, for example the lake depth in the world view, what level of data measurement was it? What about the roads? Remember what your textbook says about nominal, ordinal, interval and ratio scales, and remember also that nominal is probably the most common measurement level.

- Given what you have seen, how does the analytical paradigm differ from the communication paradigm, especially in view of the possible numbers and types of maps you can produce? Is the communication paradigm a thing of the past, or do you see it in this GIS exercise?

- What general types of map projections are available for your ArcView maps? What are the general properties maintained or affected by each?

- List some basic geographical terms used in this exercise and think about how they make you think geographically (how they modify your intellectual filter). What are some specific ArcView terms you will have to know and how do they relate to the general terminology (e.g. view versus database, theme versus coverage, etc.).

- In your examination of the databases and their properties you looked at the types of projections employed. What types of grid systems were employed in the three databases? What's the difference between a grid system and a projection?

Shutting Down

It's time to shut the computer down now. As you quit the databases and ArcView the software will ask you if you want to save any changes to your project. *PLEASE SAY NO!* We will be making some changes later on, but, because others will need to go through this exercise after you please do not change anything so that their laboratory experience will be the same as yours. Also, please listen to your laboratory instructor regarding the computers and the condition in which you are expected to leave them. Thanks for your consideration.

Laboratory Exercise 3
ArcView® Object, File, Database and Directory Structures

Introduction:

There are many file types supported by ArcView, as well as the cartographic objects themselves. In addition, ArcView has some default directory structures that will ultimately lead you astray if you are not careful. Among the more difficult tasks for the first time user of ArcView is keeping track of the object types, file types and directory structures. It is also important to understand how database structures are organized within GIS so you have a basic understanding of how you may search for and manipulate the data contained in the database.

Learning Objectives:

This exercise is designed as a follow up to exercise 2 where you learned some of the basics of getting around in ArcView, as well as some of the more basic map concepts in a digital environment. Specifically you are going to be introduced to the basic cartographic data structures (i.e. points, lines and polygons), the types of files you can use and/or produce (i.e. files with extensions such as .shp, .tiff, .apr, etc.), the basic database structure and requirements for ArcView, where ArcView will save your files, and how you can get at them easily. Beyond just the technical aspects of how to do your work with ArcView you will also be looking at the relationships between the software components and the basic concepts of how a GIS works.

When you are finished with this exercise you should be able to:

1. Identify, define, and load different kinds of data.

2. Load a new project as well as the associated shapefiles.

3. Use the help function to expand your knowledge of terminology.

4. Navigate ArcView to learn about data structures, data models, cartographic paradigms.

5. Manipulate shapefile data and save your changes as a separate project.

Linkage to your text:

In laboratory exercise 2 we covered some of the learning objectives for chapter 3. We will continue to look at some of these as well as some from chapter 4. From chapter 3 we will be looking at objective 4 which indicates the linkage between the entities and attributes through the map legend. We will also be looking at the impacts of class interval selection, symbolism, and simplification on the development of cartographic databases (# 7), examine the differences between cartographic and geographic databases (#8), and discuss some of the difficulties associated with specialty maps (#9). From chapter 4 we will examine the difference between simple graphics and maps (#1), examine the database structures available within GIS and specifically which kind is used in ArcView (#3), begin using the terminology of relational database management in your work (#4), compare the object structures (point, line, area) available in ArcView and other GIS software (#5), and examine the concept of topology (#8). There will also be mention of some of the other learning objectives, but these are the primary objectives we will focus on.

Methods:

1. Start up ArcView 3.x as in laboratory 2.

2. Click on the "new" button to create a new view. As you remember this view is a workspace for the current work session. Notice that the view has no themes; you will now add a theme.

3. Use the "add theme" button or go to the view menu and select add theme. As you do this you will notice on the bottom left corner of your screen a pull down menu that indicates your option to list files by type. This allows you to incorporate a number of additional file formats. For example, you could have satellite imagery, or other compatible file types listed there. For imagery, the following file types are available to you.

- Image data types supported by ArcView

    1. BSQ, BIL, and BIP
    2. ERDAS: LAN and GIS
    3. ERDAS IMAGINE (when the IMAGINE Image extension is loaded)
    4. JPEG (when the JPEG Image extension that comes with ArcView is loaded)
    5. BMP
    6. Run-length compressed files
    7. Sun raster files
    8. TIFF, TIFF/LZW compressed, and GeoTIFF

9. ARC/INFO grid format data can be displayed as single band images

While there are numerous file types that ArcView will support, (e.g ARC/INFO export files [.e00]) and a very large number of file types used within ArcView itself, you will not be using all of them here. Instead you will stick to the file types available in the databases with which you are working. We will revisit this to some degree later on. As you begin to open your project (qstart from lab 2, for example), look at the bottom left of your screen and examine the extension for the

file type you are opening. The .apr extension indicates that you are opening an ArcView project file, within which we have a number of additional file types (see # 4 below).

4. After having opened a new project and a new view, navigate to the esri/esridata/Canada directory and click on the rivers.shp file. Click OK to add it to your view. The **shapefile** is one of the most basic types of object file structures used in ArcView. It might be useful to examine what some of these basic file types are all about. Rather than having you try to look at all of them, instead use the help pulldown menu and, using the help index, search by topic first for "shapefiles, described." When you click on it you will receive a wealth of information (probably more than you can digest in a single sitting). You will also notice that shape files come in a number of types. Examine, for example, a point and a line. We'll look at polygons in our review questions below. Relate this information to what you have learned in your text, especially regarding dimensionality, and how these forms of the shape file are developed (e.g. a line being composed of at least two point locations). You now have the ability to search for any number of file types or for other information as you need. Now, let's continue on.

5. Now that you have selected the **rivers.shp** file, think about the probable source of this

information. Do you think it was from surveyed data or do you think is was input from an existing document? In other words is this a geographic or a cartographic database with which you are working? Add another theme such as the **province.shp**. Again ask yourself whether this is a geographic or cartographic database. Examine the tables associated with the province.shp theme. You should notice that each column has only a single value (**atom**) associated with it. This is a basic requirement of the first normal form for relational database management systems. It should also suggest that you are working with a database management system in ArcView that is linked to the graphics portion of the database. Examine what information is available in the shape file. While it indicates the area and perimeter, does it also include all of the point, line and area data as well? What does this suggest about the type of GIS with which you are working (e.g. hybrid or integrated)?

6. Imagine that, instead of provinces, each polygon is, for example, a soil polygon or a vegetation type. Think about the problems you might encounter while doing analysis with such maps.

7. Add the roads theme and examine its properties. On the left you notice a number of icons available to you. Go to the geocoding icon. What does geocoding mean? When you click on this icon you notice a new window and, on the left side you will notice a number of "checked" items that say something like "left from," "right to," etc. These are pointers indicating possible explicit relationships among the polygons. In other words, they show us that ArcView can have explicit relationships among existing entities. Although shapefiles are not topological, **topological data structures** can be built within ArcView to provide these relationships.

8. Make the Provinces theme active. Check the box next to the themes to turn them on. Notice how the software assigns colors to these themes. It seems somewhat arbitrary but you can change the colors by double clicking on the theme and then double clicking on the symbol. You can alter such things as the width and type of line used and its color. Experiment with these options before proceeding to the next step. Think about what this means in terms of the availability of output map data (circa the communication paradigm) versus the digital GIS database itself (circa the analytical or holistic paradigm).

9. Now you are going to make a quick layout. Go to the view pulldown menu and choose layout; take the default landscape template and click OK. Notice how the bar scale is empty; this is because no scale is set in the view properties. Once you have created a layout you can access it using the Layout icon on the left.

10. Click on your view window and go up to the view menu and click on properties and change the map units to miles. Now go back to your layout window and see how this affects your layout. The bar scale should have some real world units associated with it. Consider the utility of the bar scale as a method of reporting map scale within a system that allows you to change scales at will and output the map at different scales than they are shown on your monitor.

11. Notice that the name of your layout is not very descriptive. Change the name by double clicking on the name (Layout1) and rename it something more appropriate (e.g. Roads).

12. Now save your project to your workspace under the subdirectory you (or your instructor) created earlier.

13. Next copy your project to a floppy disk. Start ArcView 3 again and open the project from your floppy. Does your project come up on your monitor? Why or why not? If you only copied the file with the .apr extension the project will still load up, but why? If you took your disk to another computer that did not have the Canada data loaded would your project still work? NO! The reason is the file with the .apr extension acts as a pointer file that knows where the files to your project are located. If you just copied the .apr file to your floppy ArcView 3 will still be able to find those files on the hard drive. But if you tried this procedure on another computer without the other associated files, in this case files with .shp, .dbf, .shx, .sbn, and .sbx extensions, your project would not work.

Before you quit ArcView look at some of the review questions below.

Review and Discussion Questions.

Now that you have viewed, changed and saved the Canada database you might want to revisit it to answer some of the questions below. Remember that it is important that you look over these questions. Some of the questions are designed to get you thinking rather than to elicit specific answers so take some time to relate them to your reading and the lecture material.

1. When you viewed the database themes and looked at and modified the legend you were examining a computer version of a map. As with any map, the legend links the entities with the attributes in the final output document. However, because we are working in a computer environment, especially within the analytical cartographic paradigm, the links that exist between attributes and entities are also found elsewhere. Describe, in general terms where you find the attributes, and how they are linked to the graphical entities.

2. In your laboratory exercise you were asked to make some changes to the legend. What impact does this have on the availability of information to the reader (e.g. a client for whom you are doing the work), assuming the client wishes only to see the results of your work in map form, rather than receiving the entire database? What other forms of symbolization, generalization, and simplification did you observe from your database? For example, were all the boundaries extremely detailed, or did you detect some generalization and simplification while you observed them? What do you think were the sources of these (i.e. was this prepared as a cartograhic database or a geographic database)? What changes in abstraction take place based on whether the database was prepared as a cartographic versus a geographic

database? How might this affect the results of analysis?

3. What is the relationship between a shape file and points, lines and polygons? Where do line files inherit their properties from? Examine the help menu for line objects (under shape files) to see what information is given about what a line file is when you answer this question. Answer the same questions about points. Examine "converting a shape as polygon" in the help menu. What information is included about polygons.

4. Does ArcView have a topological data structure? How do you know? Describe what a topological data structure is.

5. Is ArcView a hybrid or an integrated GIS? What is the difference?

6. What do we mean when we say that an .apr file is a pointer file? What relevance does it have to storing and retrieving ArcView databases?

7. Given that it is very hard to measure things with a bar scale, what possible utility could it have for GIS? What other basic types of map scale are available? Describe them.

Laboratory Exercise 4
Getting Data Into ArcView®

Introduction:

So far we have been examining databases (comprised of entities and their associated attributes) that were developed by someone else. While this has been useful to allow us to understand the inner workings of a vector-based GIS in terms of organization, object, file and directory structures, etc., and while it has allowed us to examine the elements of cartography as they apply to the GIS, it does not help us understand how the databases were produced.

Most of us get involved in GIS because we want to do analysis. However, before analysis can proceed we need to produce a database that will be useful to us in our work. The production of such a GIS database involves several stages.

(1) The map must be prepared for digitizing.

(2) Control points must be identified.

(3) The coordinate system and map projection should next be identified.

(4) The map must be registered to the digitizing environment.

(5) Digitizing entities.

(6) Adding attributes.

As you might guess, there are ample ways to mess up your database at any of the stages listed above. For this reason, and because most real-world databases are typically rather large compared to the one you will be working with here, the time and costs involved in data conversion and database development can often range from 60 to 80% of that for the entire GIS operation itself (including analysis, output, etc.). This exercise will help you to understand the process, and will give you some idea of the time commitment involved for just one simple map.

Learning Objectives:

This exercise is designed to help you learn and practice the technical procedures required to produce a cartographic database from scratch within the ArcView software system. Specifically, you are going to prepare you map for input, select control points, deal with the map projection and coordinate system, register your map to the digitizer, digitize your entities and produce a simple set of attribute tables. Beyond the mechanical nature of how this is done you are going to conceptualize the input process within the framework of the GIS coordinate transformations that take place as you go from your original map document to digitizer coordinates to GIS database and back.

When you are done with this exercise you should be able to:

1. Prepare a map document for digitizing.

2. Define input projection, scale, grid system and data dictionary definitions for your input map.

3. Register your map so you can perform multiple digitizing sessions on the same map.

4. Digitize point, line, and area entities using the ArcView software.

5. Save the themes you produce in digital form.

Linkage to your text:

Given that you are going to be asked to conceptually link your digitizing process to the coordinate transformations that take place (Figure 5.3 in your text), you can plainly see how strongly this laboratory links to your text. Chapter 5 is entirely on the input subsystem and, while there are some learning objectives in your text that have less to do with this particular laboratory than others (e.g. # 11 on aerial photography input to a GIS databases), most of the learning objectives apply, to a rather high degree, to what you will do in this laboratory.

Methods:

Needed Materials:

    1. Digitizing Tablets (configured by your laboratory instructor)
    2. ArcView Software
    3. Hard copy of map document (included)
    4. Pencils (colored if possible)
    5. Xeroxed copies of original map.

Time to finish (one to two weeks)

24

Step 1: Map Preparation

Prior to performing this laboratory, you might want to read pages 137 to 144 in your textbook. The appendix of this laboratory includes 1 copy of the generalized soils map of Gove County, Kansas that you are to prepare for digitizing. **This map was reproduced from an original scale of 1:253,440. Of course the one you have is no longer at that scale. However, each of the small grid squares on the map is exactly 1 mile on a side, so you can use this for determining scale.** *Under normal circumstances as you prepare your map it might be useful to cover your map with clear mylar to provide a way to do these markups without damaging the document.* For our purposes, however, a Xerox is sufficient. You might want to make additional copies of this map.

Using a pencil (preferably a colored pencil, not a pen), place dots, asterisks, or whatever marks you prefer to determine the points along the, linear and polygonal objects which you will digitize. It does not matter which direction you move along the polygons, as long as you are consistent. It would also be useful if you would number the polygons so you know the order in which you are going to digitize. You should also number the linear objects, and point objects in the same way. The methods by which you select what you are to digitize, in which order, and how many points you use will vary for each individual. Our primary purpose is that you know beforehand where you are going to place the digitizer puck and exactly what strategy you are going to use. In short, any planned strategy is better than no strategy at all.

Because you will need to provide some coordinate information later you might find it useful to identify control points now (and label them on the copy of your prepared map). For control points you might want to use the intersections of known latitude and longitude lines such as those below:

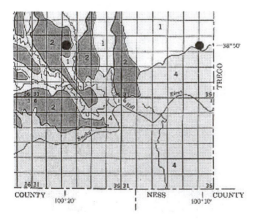

Record some additional information about your map that will be useful later on. Such information might include.

     1. Source of the data._____

2. Input scale (based on the printout at the end of the laboratory). _____

3. Input projection (if known). _____

4. Input grid system (if known). _____

5. Original material (e.g. paper or mylar) _____

6. Any other information that might be useful. _____

These data will be useful for production of a simple data dictionary and metadata for your coverage(s) or themes.

Step 2: Digitizing

The first step in the digitizing process is to set up the digitizing tablet for use. This is done in ArcView using the WinTab manager setup program that allows you to configure the buttons on your tablet's puck, one to perform a left mouse click action (to add point features or to start digitizing line or polygon features), and another to perform a left double-click action (to finish digitizing line or polygon features). You can obtain further information on this by using the online help function under 'digitizers, configuring puck buttons for use with Arc/View. You should be aware of this, although your instructors will have prepared the digitizing tablets for use prior to your actually digitizing.

Place your prepared map (including its control points and detailed instructions for input) on the configured digitizing tablet. Load the digitizer extension:

1. Make the project window active.
2. From the File menu, choose Extensions.
3. In the extensions dialog, click the check box next to the Digitizer extension.
4. Press OK.

Prepare the view into which you will digitize your map. This requires you to specify the projection used by the paper map you are about to digitize. To set the map projection for your view:

      1. From the View menu, choose Properties.
      2. In the dialog that appears, click the Projection button.
      3. In the dialog that appears, set the projection properties so they match the projection used by the paper map you'll be digitizing. If necessary, set the custom projection parameters to match your paper map's parameters. For this laboratory we will assume that we are using "projections of the world" for category and "geographic" for the projection.

      4. Press OK.
      5. Set the desired Map Units and Distance Units from the dropdown list. By default, the map units are meters. We will use decimal degrees for the map units and miles for the distance units.

As you digitize features with a known projection, the software automatically converts the coordinates and stores them in decimal degrees rather than degrees, minutes and seconds. However, if your map has no projection assigned to it, it is still possible to digitize a map under such conditions, but the coordinates you digitize in will not be converted to decimal degrees as you digitize. Instead they will remain in whatever projection the map uses (as determined by the control points you digitize). This means that you won't be able to change the projection and you won't be able to register the theme to other themes unless they too were digitized from the same map in the same projection (whatever that is).

To digitize features you need to **register** the paper map to the geographic space in your view. You do this by telling ArcView the ground coordinates for the control points you identified while preparing your map. If you want to digitize features in digitizer inches you don't need to register your map. We will be registering the map. to do this, follow these steps:

1. From the View menu choose Digitizer setup (Not there? Either you forgot to load the digitizer extension or your digitizer is not connected to your computer).

2. In the dialog box that appears, specify how much error you will allow by entering a value in the Error Limit field. The default is 0.004 inches.

3. If your view is projected, you can use the Units list in the Digitizer Setup dialog to specify what units the ground coordinates for the control points will be entered in. You can enter ground coordinates in either decimal degrees or in the units your view is currently projected into, such as meters or feet. This option is not available if your view is not projected.

4. Click the digitizer puck icon and then using your digitizing tablet, digitize the control points you identified earlier on your paper map. A record appears in the dialog box for each control point you digitize.

5. **Type** in the actual ground coordinates for the control points in the X coordinate and Y coordinate fields. As you add each control point an asterisk will appear to indicate that it is recorded.

6. Once you have entered at least four control points and their true coordinates, ArcView displays the error at each control point. It also shows the Root Mean Square (RMS) error and displays that as calculated error. The RMS is the difference between the original control points and the new control point locations calculated by the transformation process. The transformation scale indicates how much the

map being digitized will be scaled to match the real-world coordinates.

7. If the calculated error is larger than the error limit you chose, do one of the following:

    a. re-digitize the control points

    b. re-enter the corresponding ground coordinates by re-selecting the points on the screen or typing in the values again.

    c. increase the value in the error limit field so the new value is larger than the calculated error. (press the tab key to accept the new value).

8. When you are done with number 7, click the Save button to save the ground coordinates for future use.

9. Click on the Register button. This will register your map.

10. Now make the project window active. This is the window with the views, tables, charts, etc. icons on the left hand side. Make the tables icon active and click the add button. Navigate to your workspace and select your tic.txt file that you just created. Highlight it and click OK.

11. Make your view window active and go to the view menu and select add event theme. The file you want to use is your tic.txt. Click OK. This will add your tic file as a theme so you will be able to see it in your view. Thus you will be able to see if the polygons, lines and points, that you will be digitizing shortly, will be in the correct positions.

Now that you have prepared and registered your map you may begin digitizing. If you want to digitize your features into a new theme, choose New Theme from the View menu to create the new theme. You will need to do that with your soil map because it includes points, lines and areas.

At this point your lab instructor will show you how to digitize (in absolute mode), how to create separate themes, and how to save your data.

Products:

The following should be turned in to your laboratory instructor as a final product for this and the previous two laboratory exercises:

1. For part one of laboratory exercise 4 submit a Xeroxed copy of your prepared map with your name at the top.

2. For part two of laboratory exercise 4 submit a 3.5" floppy with your prepared database. It should have a label on the floppy with your first and last names (printed legibly), and lab report #1.

3. For the final part of your laboratory report, answer the following questions.

    1. Why is it important to prepare a map before digitizing it? What problems can result from not preparing a map prior to digitizing?
    2. What problem can result if you do not have a map projection on the map you are digitizing?
    3. When you set of your digitizing tablet, one option you had was to set up stream mode digitizing? What is this?
    4. Are there any advantages or disadvantages to this mode over points mode digitizing?
    5. What decisions had to be made to select points to be digitized? How does this relate to the idea of sampling?
    6. What is RMS Error? What impact does this have on the accuracy of your map?
    7. What is fuzzy tolerance? How do you set this up with ArcView?
    8. What is map registration? Why do you need to do this?
    9. Describe the sequence of map transformations you went through to produce your digital database. What additional transformations will you need to go through to produce map output?
    10. What type of database (cartographic or geographic) are you producing through digitizing this existing map? How does the accuracy of this type of database compare to the alternative?

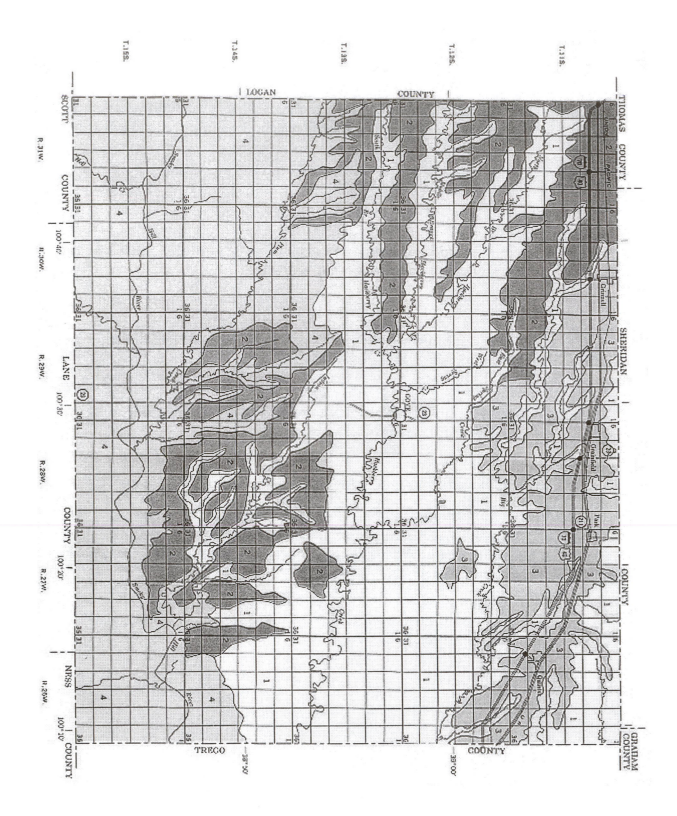

Laboratory Exercise 5
Adding Attributes and Editing Your Data

Introduction:

As Fundamentals of Geographic Information Systems (chapter 6) indicates, there are three basic types of error that can occur -- (1) entity errors, (2) attribute errors, and (3) entity-attribute agreement errors. While some texts indicate that there are only two types (i.e. #'s 1 and 3 above), this is not always the case because misspellings of attributes means that, while the linkage between entities and attributes is correct, the GIS will not be able to establish this linkage and retrieve the correct entity, because it's expected attribute doesn't correspond to the spelling of the key being searched. In some cases it is often advisable to establish simple attribute codes rather than using complex terms for searches.

In laboratory exercise 4 you spent much of your time with the input subsystem of the GIS by digitizing an analog soils map of Gove County, Kansas. Although some editing was, of necessity, performed during that process, it is safe to assume that there are still mistakes… mistakes that will now be corrected. In addition, while you were digitizing the point, line and area entities of the soil map you did not input any of the entities. With many simple raster systems (especially those that are not linked to a database management system) the attributes are necessarily entered at the same time as the entities because the entities are effectively created by the order in which the attributes are input as a matrix of values. For the most part, with vector systems, the attributes are added after the entities have been encoded. This way you can focus on the entities first to be sure you encode them correctly, then add the entities as a separate operation, again so you can focus on them.

Learning Objectives:

This laboratory will give you an opportunity to look at the capabilities of ArcView for editing both entities and attributes. The entity editing will be limited by the shape file structure used to represent the data. As such, you will primarily be working with vertices to edit the shapes you have created. The vertex is the point where two sides of a shape meet. This data structure allows you to add, delete and move vertices for a given shape -- an action that will also change the neighboring shapes accordingly. This gives you much more flexibility and ease of editing thanC/INFO, where moving lines for one object could conceivably adversely affect adjacent cartographic objects. In this exercise you will see that, while digitizing in ArcView can be a major undertaking, the editing of entities is actually quite simple and straightforward.

Beyond editing the entities themselves, you will edit the contents of the attribute tables by adding relevant data for each theme table. This will demonstrate to you that, should additional data become available for the point, line and area shapes in your database, you can add them easily. Or, if you find that the data you have entered are incorrect, you can delete them and enter new attributes at will.

When you are finished with this exercise you will be able to:

1. Identify entity errors and correct them using ArcView software to physically move vertices to the correct locations.

2. Enter and/or edit attribute data and identify the linkages between the attributes and their entities.

3. Recognize the differences among the three types of GIS error.

Linkage to your text:

This laboratory corresponds to both chapters 5 and 6 in your text. Because we spent laboratory 4 on the learning objectives for chapter 5, we'll concentrate on the learning objectives for chapter 6. Although the shape file structure is more user friendly than many of its counterparts we can still identify (at least after having performed laboratory 4) with learning objectives 2, 3 and 4 dealing with the types, identification and correction of error during database development. Learning objectives 1 (dealing with tiling), 6 (having to do with edge matching), 7 (conflation) and 8 (templating) will be more frequently encountered when you are working with large databases and where the databases need to be co-registered, so we won't deal with them directly in the exercise, but they certainly are fair game for review questions. Learning objective 5 (dealing with projection changes) is relevant primarily for laboratory exercise 4, especially with regard to the limitations of analog maps without known projections.

Methods:

We will begin by adding attributes to the Gove County Database you digitized in laboratory 4. You are to add attributes for all entities you digitized (point, line and area shapes). In ArcView tables you can change the cell values and add and delete both records and fields. You can also perform mathematical, logical, and text operations on existing fields and store the results in a new field (but we will save this for a later exercise).

1. From the File menu, open your Gove County database.

2. You will begin by editing the point data so get into your "towns" (point) view.

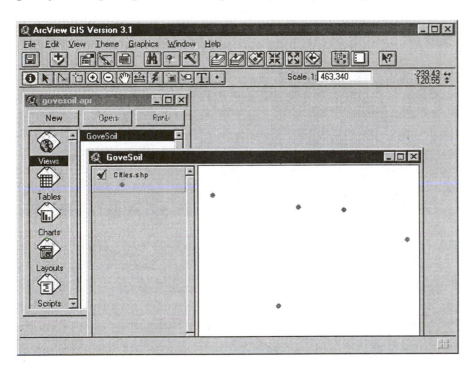

3. To add attributes (by editing fields) to your database you must turn on Table Editing.

4. Make sure the theme table is active. From the Table menu, select Start Editing (The field names become non-italic, confirming that the table is editable). **Do not edit any of the existing**

fields because they were generated by your digitizing process. Editing these fields will compromise the integrity of your database.

5. From the Edit menu, choose Add Field to display the Field Definition dialog box.

6. In the Name input box, highlight the default name, "NewField," and type **Towns**.

7. In the Type input box, highlight the default type "Number" and type "Text."

8. In the Width box, highlight the default width and insert a number that corresponds to the largest text you have representing the towns in Gove County.

9. Click OK and add the new field to the table.

10. Now enter the values you want for the towns.

11. By highlighting the remainder of the fields, add streams, soil types, roads, and railroads as necessary.

Now that you have entered attribute data into your database it is time to see how the entities themselves are edited. We'll begin with point data.

1. Within the Gove County database make the towns.shp theme the active theme by clicking on it in the table of contents. From the Theme menu, select Start Editing. Use the pointer tool to select the point you want to delete or move. To select more than one point at a time, hold down the Shift key as you click on the points, or drag over them. Selection handles will appear around the selected feature(s).

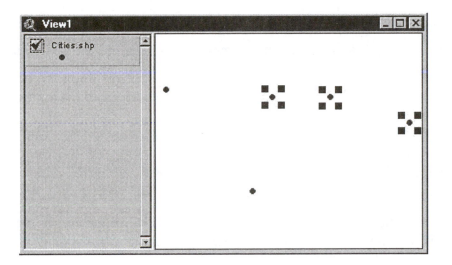

2. If you want to delete the selected point, press the Delete key on the keyboard. The point is removed from the view, and the corresponding record in the attribute table is automatically removed (topology remember?). To move the selected point, drag it to the new location.

3. If you accidentally delete the wrong point, select Undo Edit from the Edit menu.

4. When you are done, choose Stop Editing from the Theme menu. Choose NO (unless you really don't have your points in the correct locations) when you are asked whether you want to save changes. Given that this was just an example exercise you don't want to corrupt your existing database.

Now we'll move on to editing line data.

1. Within the Gove County database make the rivers.shp theme the active theme. You remember that when you digitized this particularly line theme you had to set your snap distance (fuzzy tolerance). This tolerance can give you some difficulty while digitizing, so you might be able to edit some of the error now. From the Theme menu, select Start Editing. Use the Vertex Edit tool (second row of the button bar) to select the lines you want to work with.
Experiment with vertex editing using the rivers theme. You may want to experiment with the other line themes as well. As you make changes, check the tables to see what has happened. When you are finished (again, unless your edits were really useful rather than just for practice),

do not save your work. Check out the information below to get a feel for what you can do with the vertex edit tool for line entities.

2. When you edit a line, you can choose to preserve topology, or not by the way in which you select the feature:

- When you select a single line, any edits you make to the vertices will only affect that line.

- When you select a line segment common to two lines, any edits you make to the vertices will affect both lines.

- when you select a node common to two or more lines any edits to that node will affect all the lines that contain that node.

3. To move a vertex with the Vertex Edit tool.

- Place the cursor on the vertex you want to move. When the cursor appears as a crosshair, hold down the left mouse button and drag the vertex to the new position.

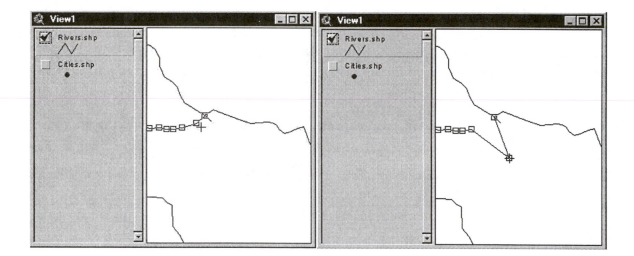

4. To add a new vertex with the Vertex Edit tool.

- Move the cursor to the position on the line where you want the new vertex. When the cursor appears as a target, click the left mouse button.

5. To move a vertex with the Vertex Exit tool.

   • Place the cursor on the vertex you want to delete. When the cursor appears as a crosshair, press the Delete key on your keyboard.

6. To reshape a single line.

   • Click the Vertex Edit tool.

   • Click on the line. A hollow, square vertex handle appears at each vertex of the line.

   • Now when you move, add, or delete vertices, only the single line will change.

7. To reshape a segment common to two lines

   • Click the Vertex Edit tool

   • Click on the common line segment. Square vertex handles will appear at each vertex of the shared segment, and round anchors will appear at the vertices at each end of the common segment.

   • Now when you move, add, or delete vertices, both lines will change.

8. And finally, to edit area (polygon) data.

   1. You are going to practice editing polygons using the soils data you generated earlier. As before, unless you are really making actual edits you wish to keep **do not** save your work. Within the Gove County database make the soils.shp theme the active theme by clicking on it in the table of contents. Using the Vertex Edit tool to reshape a polygon by moving, adding, or deleting vertices. When you edit a polygon, just as when you were working with line data, you can choose to preserve topology, or not, by the way in which you select the feature:

      1. When you select a single polygon, any edits you make to the vertices will affect only that single polygon.
      2. When you select a shared boundary between two polygons, any edits you make to the vertices will affect both polygons.
      3. When you select a node common to two or more polygons, any edits to that node will affect all polygons that contain that node.

2. To reshape a single polygon.

- Click the Vertex Exit tool.

- Click inside the polygon. A hollow, square vertex handle appears at each vertex of the polygon's boundary.

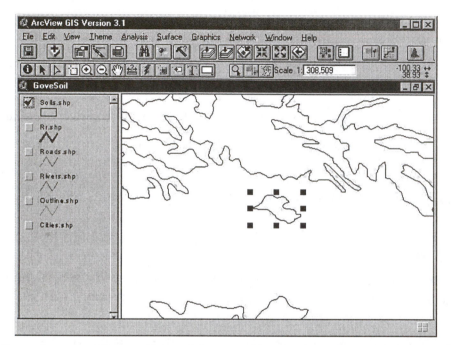

- Now when you move, add or delete vertices, only the single polygon will change.

3. To reshape a common boundary between two polygons.

- Click the Vertex Exit tool.

- Click on the common boundary. Square vertex handles will appear at each vertex of the shared boundary and round anchors will appear at the vertices at each end of the common boundary.

- Now when you move, add or delete a vertex, both the polygons will change.

.

4. To move a node that is common to a number of polygons.

1. Click on the Vertex Edit tool.

2. Click on the node that is common to two or more polygon features. A square vertex handle will appear at this node, and round anchors will appear on the next closest vertex on each polygon.

3. When you move the common node, all polygons that share this node will change.

Products:

This laboratory exercise is designed to give you a chance to practice. Spend sufficient time on this that you become familiar with what the editing tools are like. Look, for example at what happens to the data in your tables as you make changes. For example, see what happens to the attribute data associated with the area of each polygon as you change joining polygons with the edit vertex tool.

Review Questions:

1. How do you know that you have selected a point entity feature for editing? A line entity? A polygon? Multiple polygons?

2. What happens to the adjacent entities if you move the vertices of a polygon that is connected to another?

3. What impact does moving polygon vertices have on the tabular data?

4. What are the three major types of error associated with digitizing?

5. How does ArcView assist you in correcting each individual type of error you outlined in question number 4?

6. Which types of errors are hardest to find and correct?

7. Which types of errors does ArcView not prove as helpful in correcting?

8. Describe how you add attribute data to entities.

Laboratory Exercise 6
Finding and Locating Objects

Introduction:

Within the analysis subsystem of the GIS among the simplest, but most common tasks employed in real-world GIS applications is that of inventory control. Many people use maps to archive important information and are often called upon to simply retrieve this information whenever a request is made. For example, a natural resources department may be called upon many times to determine whether or not there are endangered species located within a proposed construction site. In the analog domain, the department personnel typically pull out the traditional map document that is normally stored in a large, cumbersome map cabinet, and then identifies the specific location and determines whether or not an endangered species happens to be located within the specified land parcel. For each request (and there are frequently many), the procedure must be repeated. This process is time consuming, and also requires the space required to store the large map cabinets. In addition, because the process of producing new maps by hand is time-consuming, the maps may be out of date and in error.

As you can see from this example, and there are hundreds more you could probably think of, the utility of the GIS as a storage and retrieval device often exceeds the actual analysis. Because the process of selecting individual pieces of data is more closely associated with the process of map reading and interpretation, I have included it within the analysis subsystem. Finding objects can entail simple searches through the database for either entities or attributes (more often the case), it can entail identifying absolute or relative locations (more on relative location later), or it can entail combined search strategies that require the GIS operator to construct a set of search criteria most often based on some aspects of the attributes themselves.

When you go through this laboratory keep in mind that, while the analyses you are performing may seem rather mundane, such approaches are often important and should not be downplayed. Beyond the relative frequency with which such requests will be made, their utilitarian importance and their importance as building blocks of more complex analytical procedures later on will become clearer as you continue on with the laboratory exercises. In fact, most cartographic models rely quite heavily on the simple accounting procedures you will experience here.

As you go through the laboratory try to envision scenarios where you are likely to be employed as a GIS professional by an organization that could best be described as a neophyte when in comes to this technology. Think about the types of simple requests such an organization might be expected to make of you in the early years of their use of GIS. For example, anticipate what simple requests some of the following organizations might make of you: (1) a major metropolitan newspaper, (2) a real estate company, (3) a utility company, (4) a conservation organization. When you are done, keep these in mind for the remainder of the course, especially as you learn more about what a GIS can do. Then, develop ideas about how the analytical power of the GIS could be used to further enhance the operations of these organizations beyond the simple accounting procedures they may initially request. This will give you an opportunity to enumerate your potential as a future employee.

Learning Objectives:

This laboratory exercise will give you an opportunity to explore the methods ArcView has available for finding and locating objects. You will learn the techniques for isolating objects based on attributes and how to find attributes by isolating the entities in a graphical environment.

When you are finished with this laboratory you should be able to:

1. Select point, line, and polygon entities by pointing to them on the map with the identify tool.

2. Select point, line, and polygon entities within a window.

3. Examine the tabular attributes of the selected objects.

4. Promote the tabular attributes of the selected objects so they are all on top op your tables.

5. Determine the absolute coordinate locations of the objects you selected.

6. Use the Query tool to select objects based on attribute criteria.

7. Suggest combinations of the query builder tools to make more complex queries.

Linkage to your text:

Most of the learning objectives from chapter 7 (Elementary Spatial Analysis) are employed in this exercise. The textbook learning objectives most strongly emphasized, through your laboratory experiences are the following: understand why analysis is important for database builders (#2), explain the process (within your software) for isolating, identifying, counting, and separately tabulating and displaying individual items and describe the differences in process for each (#'s 3 & 4), explain measurable attributes for lines and areas (#5).

Methods:

Selecting map features in a view:

1. Copy your Gove County Soils database and place it in your class subdirectory. Add the soils, roads, cities, rivers, and railroad themes.

2. Make "Cities" the active theme.

3. Using the Identify tool, click on each of the towns you digitized earlier to obtain the tabular information for each.

4. As you move from point to point in the database note how the location indicator (upper right) changes as you move. This gives you an idea of the absolute location of each feature in the database.

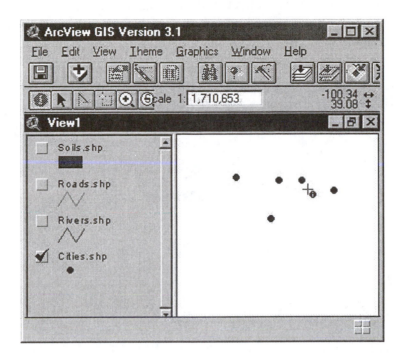

5. Make the "soils" the active theme.

6. Repeat steps 3 and 4 above for soils. Note the information that you manually included and what the software included in the tables. It will identify such attributes as the shape type, and soil type.

7. Repeat step 6 for your rivers, roads and railroads.

8. Now that you have identified the features it is time to select specific ones for viewing. We'll use the line segments from the streams you digitized. Close the Identify tool and click on the Select Feature tool . Holding down the Shift key, click on each of the line segments for one of the streams you digitized. Note how the software highlights the selected segments in yellow. **NOTE: if you select an incorrect feature you can unselect it by holding down the Shift key and clicking on it again.** Try this out.

9. You have selected the entities, but the attributes are also selected at the same time. Click the Open Theme Table button on the View button bar. The theme's attributes table opens up. Scroll through the attribute table to note the items that are highlighted in yellow. These correspond to the attributes that you selected in the map.

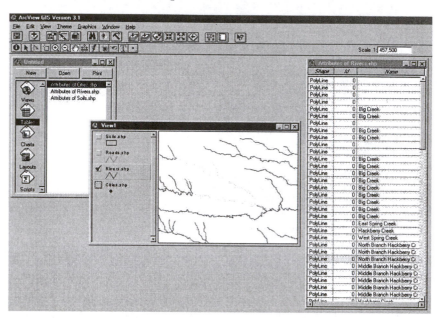

10. Click the Promote button on the Table button bar. The stream segment attributes for the line segments you selected are now placed on the top of the table for easy viewing.

11. Click on the table's title bar (in blue), then drag it to the upper left of the ArcView window. Now click on the view's title bar and drag it to the lower right corner and make its window smaller.

12. Make the table active, click on the Identify tool in the Table tool bar, then click on any one of the highlighted records (let's say there is a particular segment of the stream you want to isolate, perhaps a highly meandering segment that might be causing excessive erosion). The Identify Results dialog box displays and the feature you picked flashes in the view (If you don't see the feature flash, click on the record again).

13. Move the Identify Results dialog box out of the way so you can see the view. Click on another meander segment in the highlighted records. Note how this segment is added to the Identify Results dialog box and flashes in the view.

14. These segments can now be isolated for display to others interested in stream erosion in Gove County. Close the Identify Results dialog box, then the theme table. Drag the upper left corner of the View window to enlarge it.

15. Click the Clear Selected Features button to clear the current selection. The previously highlighted features are no longer highlighted.

16. Another way to select features graphically is to use the Draw tool to draw a shape in the view, then use the Select Features Using Graphic button to select the features under the shape. Any shape that touches this shape will be highlighted. Try this out and see what you get.

Because ArcView links entities and attributes in a table, you can select features by entering an attribute value or by writing a statement, called a *query*, that specifies one or more attributes and the values you are interested in seeing. ArcView searches the attribute table for records that match your request. When you use the Find tool, ArcView finds the first feature that matches your request; when you use the Query Builder, ArcView finds all the features that match your request.

1. Make the "soils" theme the active theme.

2. Click the Find button on the View button bar, then type the name of one of your soils types in the text box that displays. Click OK.

3. Note how ArcView highlights that soil type in your coverage. Click the Open Theme Table button, then the Promote button. Notice that a single record matching your search is found.

4. Close the theme table then click the Query Builder button on the View button bar to display the Query Builder dialog box. The theme table name is displayed on the dialog box.

5. The Query Builder dialog box contains a list of attribute fields (left), a set of operators (center), and a list of attribute values (right). When you click on a field in the Fields list, all the unique values for that field display in the Values list, as long as the Update Values option is checked (default). To build a query, you double-click on a field, click (or double-click) on an operator, then double-click on a value. As you build the query, it displays in the query text box in the lower left corner of the dialog box. You can also type your query directly in the query text box.

6. In the query Builder dialog box, scroll down to the fields list, then click on "soil association" or whatever you called your soils class fields when you built your database. The various soils types available to you will appear in the Values column.

7. Now double-click on the "soil association" to place it in the query text box.

8. Click the "=" button. It displays in the query text box. Then in the Values list, double-click on any soil type you want to find.

9. Let's say you want to select two soil types. Click the "or" button, double-click "soil association" again in the Fields list, click the "=" button, then double click on another of the soil types you wish to include in your query.

10. Click the New Set button to select the two soil types you were interested in. ArcView highlights them in the view. (You may need to move the Query Builder dialog box so you can see the view).

11. When you are finished with these simple queries, look at the questions below. You might try adding some attributes (Make them up or search for them in atlases, etc. For example, find the population of the various towns by looking in a road atlas, look at the soil survey for additional data on the soils, etc.)

Review Questions:

1. Suggest what types of questions might be answerable by selecting specific attributes associated with the towns, roads, streams, railroads, and soiltypes included in your database. For example, could you add the population data for the towns, the road types, stream flows, soils capabilites, etc. to your database?

2. If you could add the types of data suggested above, what kinds of questions could be answered with simple queries of the database?

3. Assuming you are working for a Natural Resources Conservation Service (NRCS) office and someone wants to know whether or not they could use a particular piece of land for farming, beyond finding the tabular data (or including it in the database), what other questions might be asked that may require more than simple searches for attributes, or may require additional data? For example, how could you consider such information as land ownership (if you had it), with soils attributes and railroad easements?

4. Consider the use of measurements as attributes for your data. For example, how could stream segment/land distance be used to isolate highly meandering stream segments (and therefore, more erosion and deposition)?

5. How could the amount of soils of a particular association or type be used to determine viability for large-scale grain cropping?

Products:

The questions above should be turned in after this laboratory exercise. They are designed to make you think ahead and to consider both the utility of attribute search and its limitations. Be complete in your answers rather than giving a cursory answer.

Laboratory Exercise 7
Simple Measurement

Introduction:

Another simple, but quite useful capability of the GIS analysis subsystem is its ability to measure objects once they are included in the database. The measurements are based on ground measurement units in the system of measure you assign (e.g. English or Metric) when you digitize a map, but can be changed to other types of measurement as well. The diagram from your book (Figure 5.5) illustrates the transformations that take place during digitizing through analysis, and output. The analysis is actually performed (in most GIS software) in geographic coordinates.

The measurement of objects is quite simple in some systems. In fact, this is one case where ARC/INFO is easier to use than its little cousin ArcView. This is primarily because, as you digitize in ARC/INFO, the software carries on a constant calculation of line segment lengths, polygon areas, etc., and then stores them explicitly in its tables. ArcView's data structure, for some reason, does not. This limits our ability to search for lines and polygons based on their sizes. We could, however, measure these objects and then insert the appropriate dimensional information into tables for later queries. That's not where we are going, however. In this laboratory we will perform some measurements on the Gove County, Kansas database.

Learning Objectives:

This laboratory exercise will give you an opportunity to explore the capabilities of ArcView for measuring lengths and areas. You will also measure perimeters (an artifact of length) and then perform a simple ratio of perimeter / area. In addition, you will perform a simple measure of sinuosity of one of the streams. These will provide you with enough to at least get you thinking about how you could combine these operations to get more sophisticated results.

When you are finished with this laboratory you should be able to:

1. Measure simple length of linear objects.

2. Measure perimeter and area of polygonal objects.

3. Measure the major and minor axes of polygonal objects.

4. Measure sinuosity of linear objects and compare straight-line to sinuous lengths.

5. Perform a simple perimeter to area ratio.

6. Suggest ways of applying these measurements to more complex map measurements.

Linkage to your text:

Most of the learning objectives from chapter 8 (Measurement) are employed in this exercise with the exception of 5 and 6 which deal with functional distance and friction surfaces. We will revisit these when we look at Spatial Analyst later. Spatial integrity and boundary configuration (Text objective 4) will be discussed, but you will not be required to perform these operations until we work with Spatial Analyst.

Methods:

Measuring Distance:

1. Copy your Gove County Soils database and place it in your class subdirectory. You may remember that all of the themes in this database were stored in Decimal Degrees.

2. Make "roads" the active theme.

3. Select Properties from the View menu to open the View Properties dialog box. Having done that, click on the Map Units down arrow, then choose "decimal degrees" from the list. This indicates that all the data in the current view are stored in decimal degrees. To specify the units ArcView will use to report measurements, you'll set the distance units.

4. Click on the Distance Units down arrow, then choose "feet." Before returning to the view, you decide to give it a more descriptive name.

5. Click in the Name field and highlight "View1." Change the name to Road Log to indicate your interest in road miles. Click OK to apply your settings to the view. Now the view has a new name and the map scale displays in the Scale box at the upper right.

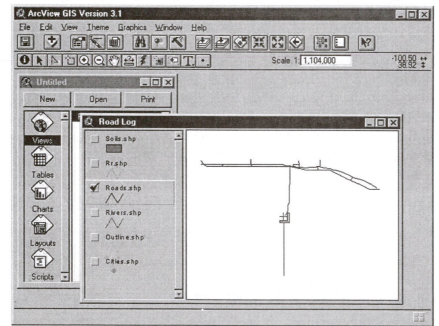

6. Now you'll use the Measure tool to determine the distance (in feet) along any of the roads. Click on the Measure tool (the cursor changes to a ruler), then click on the start of your road. Move the cursor to the end of a straight line segment of the road. Notice that ArcView draws a line segment along that part of the road and reports the length of this line in the status bar (at the bottom of the ArcView window). With the cursor directly over the end of your road segment, double click the mouse button to end the line. ArcView reports the measurement in feet.

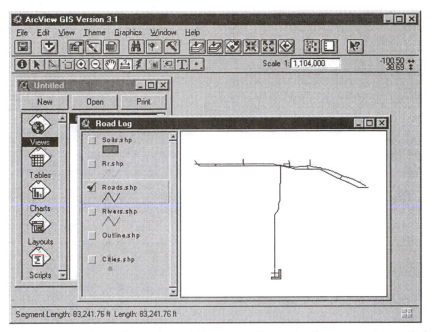

7. ArcView reports two values, Length and Segment Length. Segment Length is the length of the current line segment and the Length is the total length of all segments. Given that you have only measured one segment along the road, these are identical.

8. Continue measuring the road segments until you finish. For example, you should measure non-linear roads one segment at a time using a single measurement for each linear segment. Note the different values for Length and Segment Length.

9. Now let's measure something considerably more curved than the roads... let's measure one of the rivers (you choose). Create a new view (View2) and put the outline and rivers themes into it. But, before we begin (and of course you should now make the rivers the active theme), we want to do this in meters because most science is done using the metric system. Click on the Distance Down arrow and choose meters.

10. Just as before we want to rename the view to be more meaningful. Click in the Name field and highlight "View2." Change the name to River Geomorphology to indicate your interest in measuring river length and sinuosity. Click OK to apply your settings to the view. Now the view has a new name and the map scale displays in the Scale box in the upper right.

11. Now, using your measurement tool measure your river from beginning to end using two different lengths: first using small, incremental distances along the curved river, and second, a straight line distance from beginning to end. Record both of the values for Length (not segment length). Create a ratio of the straight-line distance (numerator) and the actual linear distance (denominator). What do you get? What does it mean?

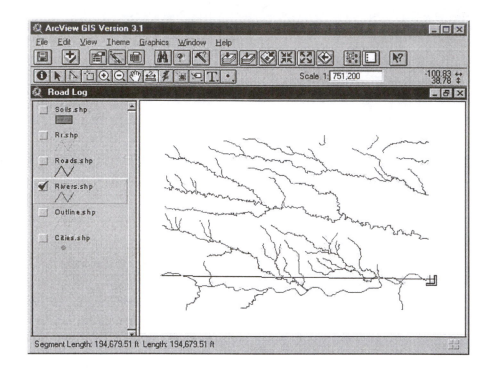

Measuring polygonal shapes:

12. Create a new theme called soils containing the soils and outline themes. Make the "soils" the active theme.

13. Select one of the soils polygons that is completely contained in your coverage. Now, using the measurement tool you used previously measure the following for this polygon in meters (be sure to record these values on a piece of paper):

    1.   perimeter

    2.   major axis

    3.   minor axis

14. Measure the area of the same soil polygon. Record this number. This can be accomplished in two separate ways.

    1.   Use the vertex edit tool and select a polygon (remember to go to the pulldown menu to select "theme" "start editing" before selecting the edit tool) and it will read the area of the polygon. You need to be careful with this one because you might corrupt your coverage by

moving the vertices around. If you do this, simply do not save your work as you exit the program.

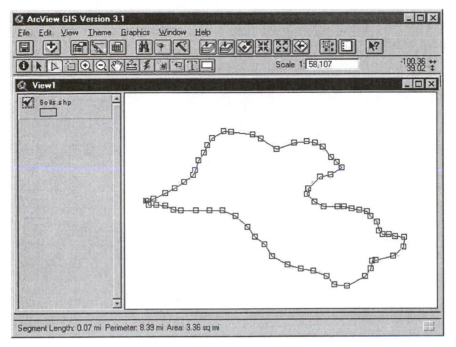

2. Use the drawing tools and select the irregular polygon (the icon looks like an irregular

polygon) and, as carefully as you can, trace over the area, double clicking when you reach the end. The disadvantage of this approach is that your measurement will be less than totally accurate.

15. Now that you know the perimeter and the area of the soil polygon, do a ratio of the two values.

       Why would you want to do this? _____

       What does it tell us anyway? _____

16. Now that you are finished measuring simple objects you may continue to experiment, or you may proceed to the questions.

Review Questions:

1. Suggest ways that you might be able to derive a more detailed measure of stream sinuosity as described on page 211 in your text.

2. While you were measuring your road, we assumed that the road was flat. Let us assume that the road was actually along a hilly surface. What would this do to your values for road length?

3. Road maps assume flat roads, but your automobile has to travel on hilly (non-flat) road segments. Assuming you had a USGS topographic map (7.5" quadrangle) and you were working for the Department of Transportation, what adjustment would you have to make to the road log values if you first digitized them from the topographic map?

4. Suppose, just for the sake of argument that you know the beginning and the ending elevation of your road segment. Suggest how you would calculate the road miles (actual road log) given your measured straight-line distance.

5. What might the major and minor axis tell us about the soil polygon?

6. How would an ecologist, or a geomorphologist use this information?

7. What does the perimeter / area ratio tell us about the shape of our soil polygon?

8. What happens to your perimeter / area ratio as your polygon goes from a circle to a more wrinkled shape (like an amoeba)?

9. Suggest how you could perform some simple form of contiguity measure with ArcView (without using Spatial Analysit).

Products:

Hand in the answers to the questions from laboratory 7. Be sure to be as complete as possible and include any references you might have especially when you are asked to speculate upon the applicability of these measurement methods by selected disciplines.

Laboratory Exercise 8
Classification

Introduction:

Because many polygonal-based GIS packages were designed to handle land use and land cover data (i.e. natural resources data), much of the time the GIS user is forced to contend with the problem of classification itself. As your text points out, this problem has been around for quite some time and the problems are not going to go away any time soon. Among the more sticky problems we find is the difference between such terms as land use (implying actual human intervention) and land cover (implying a simple classification of what can be observed on the ground). Unfortunately, much of our classification terminology is biased by our experience so when we see a number of evenly spaced trees (the land cover) we quickly come to the conclusion that this is an orchard (the land use).

While the latter problem is one with which we will struggle for years, and you will likely encounter often in your future GIS work, there also exists a less subtle, but no less annoying problem concerning classification. This problem is often more pervasive in its ultimate impact on the utility of a GIS database and involves the relationship between a classification and the use for which the GIS database is to be put in the first place. This is clearly illustrated in the analog vegetation map research of A.W. Küchler (see chapter 9 in your text for more information) in southeastern Mount Desert Island, Maine. Selecting a small area he painstakingly surveyed the vegetation of this location, collecting detailed data on herbaceous versus woody vegetation, percentage cover, tree heights, species composition and the like. When he was finished he was able to produce three distinct maps of the vegetation of the region. So, his question was, what is the relationship between each classified map and the potential uses to which each could be put?

The dilemma he faced in the 1950's is even more common today because we now have the capability to produce a myriad of maps from our digital databases. To illustrate this point, you are going to be using a digital version of Küchler's original field notes. And, you will recreate some similar inquiries by answering some of the questions he was asking about which map is better for which task. Here, however, because you are not restricted to the communication paradigm you can create maps specifically designed for the selected use.

Learning Objectives:

This laboratory exercise will give you an opportunity to explore the capabilities that Arc/View has for classifying polygonal data based on specific user needs. You will be using parts of a database created in ARC/INFO from Küchler's original field notes. This will allow you to see how a single set of collected data can have literally thousands of different meanings based on how they are separately displayed or on how they are grouped. Your laboratory is based on the 1991 paper on classification and purpose in automated vegetation maps published in the Geographical Review (DeMers 1991 – see reference in Fundamentals of Geographic Information Systems). You may want to take a look at this article to gain more insights into what this exercise is all about.

When you are finished with this laboratory you should be able to:

1. Produce both simple and complex queries of a database using the query builder tool.

2. Select criteria and weighting factors as classification tools for your database.

3. Create new database fields that enhance the utility of the database.

4. Identify the query limitations of a database that is not in first normal form.

5. Reclassify your map based on neighborhood values.

6. Reclassify your map using a buffer function.

Linkage to your text:

The text learning objectives you will be addressing include understanding the enhanced importance of classification available with GIS (#1), learning how to aggregate data (in vector) (#2), and understand the concept of neighborhoods through direct manipulation (#'s 5 & 6). We will begin looking at buffers here, and later, in laboratory 9 we will begin looking more closely at buffers and their creation (learning objective #8). We will also revisit the concept of buffering when we are introduced to Spatial Analyst. Learning objectives 3 and 4 deal mostly, although not exclusively with raster GIS and will not be covered in this laboratory, but will be covered using Spatial Analyst.

Methods:

Classification:

1. In today's lab you will be working with an ARC/INFO database named Mainelab that has been imported into ArcView. This database has been copied for you and is found in your course directory, and contains data about the plant species, woody vegetation and herbaceous vegetation found in Mount Desert Island, Maine.

2. Open ArcView and add the Mainelab database as a theme to a new view.

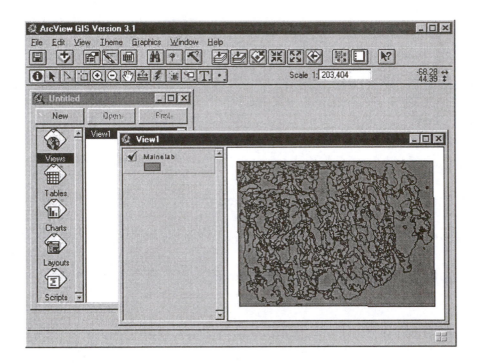

3. Wander through the tables for the species. These are the fields (columns) labeled S1-S29. You will note that there are many columns of the table that have missing data (represented by 0's). The codes should also show seven digits. Randomly select one of these seven digit codes. Write this value down on a piece of paper. Now, using the data dictionary in the appendix at the end of this laboratory, determine the genus, species, coverage, and sociability.

Mainelab-id	S1	S2	S3	S4	S5
222	0010111	0020211	0060213	0040113	037012
220	0010111	0210331	0930311	0950411	105013
214	0060213	0240165	0269912	0290111	077011
213	0020231	0190811	0210221	0210341	035013
212	0020211	0210322	0350144	1050132	107012
195	0190411	0110111	0260524	0261013	026132
344	0020221	0060121	0080131	0210211	035016
168	0020221	0150122	0210221	0210351	035013
12	0	0	0	0	0
170	0020211	0050323	0210361	0350133	041011

4. One of Küchler's suggested tasks was to examine the potential of one or more types of vegetation maps to assist the Internal Revenue Service (IRS) in evaluating the study area to determine the value of the timber holdings. To do this you would first use your query tool to select the polygons that contain commercially viable species such as white pine (*Pinus strobus*){species code = 09504}, red pine (*Pinus resinosa*) {species code = 09502}, white spruce (*Picea glauca*) {species code = 09301}, red spruce (*Picea rubens*) {species code = 09303}, and hemlock (*Tsuga canadensis*) {species code = 13201}.

5. For the IRS evaluation you will need more than just the commercially viable species. The trees will need to cover at least 75% of each area and grow as a continuous-growth extensive sheet. These last two factors are the last two digits of the seven digitit species code. So, for example if you are looking for white pine covering at least 75% of the area as a continuous-growth extensive sheet the code you would be looking for would be 0950465, for red pine it would be 0930365, white spruce would be 0930165, red spruce would be 0930365, and hemlock would be 1320165; where the last two digits indicate the percentage growth and sociability codes. Because there are up to 29 species you will have to search each of these fields (columns)

to create your map. As you use your query builder you should note that the available codes for each field show up in a window to the right. By clicking on these, the values that you select on the right are transferred to the query you are building. This will save an enormous amount of typing as you do your searches. In fact, you will discover that none of the values listed above actually occurs in the database. Try looking for a few of these using just S1 and S2. You can, of course continue to search all 29 fields if you wish to verify the absence of these codes. (Note

also that these numbers have quotation marks around them. This indicates that the database is treating these as character strings not numbers. This was necessary because of all the zeros in the coding scheme).

6. The attempted query in number 5 above shows you two things that are important in database design later on in your course. First, because of the enormity of the database it is not easy to create a set of tables that conform to first normal form, particularly where the tables are expected to be symmetrical (instead of having all the missing values coded as 0). This creates an awkward search strategy as you should have already determined by your initial searches. Second, while one or more of the cartographic documents Küchler produced may seem to provide the appropriate solution to his IRS problem, your quick examination of the database shows that there are really no commercially viable forests in the study area. In other words, either the study area or the criteria for selection of variables were not wisely selected. Before this exercise is finished you will become painfully aware of these limitations.

7. Now, given that you found nothing to meet your criteria above, try reducing your constraints a little by searching for the same tree species, but lowering your height requirement to trees that occur singly rather than in continuous sheets. Repeat your search using your query tool based on these revised criteria. Produce a map of commercially viable species.

8. Now, relying simply on your query tool to perform classification and reclassification devise a method to rank your forests into two or three ordinal groups (high, medium, low). You can do this by selecting from the set of data you already have and querying the woody vegetation codes (W1-W7) for additional criteria. At this point you will need to be creative in how you select your numbers. The only non-coniferous tree is the hemlock. You could search for tall, medium, short, shrub and dwarf evergreens first, then look for the same with hemlock later on.

9. For our next task we are going to perform a scientific analysis of the study area by determining a measure of species richness (number of species).

10. Now, with "Maine" still the active theme, get into the theme tables and get into edit mode. Now you are going to add a field to the table (in this case because what we want to do is not readily performed in ArcView). The field you are going to add is "Species #" which will be a simple count of the number of columns (S1..S29 each contains). So the maximum number of species will be 29 for any given polygon. (Note: this can take a while for all 600 polygons so you might want to team up with your fellow students to put together portions of the list and then share your numbers. Or, your instructor may allow you to limit your approach to a small subset of the database).

11. After you have created and filled in the number of species per polygon, create a map of these data.

12. Now we'll use another portion of our database that provides us more detail about woody vegetation that is not included in the species database. Our purpose here is to provide a map that is likely to show us a spectacular display of leaves as autumn approaches the coast of Maine. This will allow the guides a better idea of where to bring visitors to Acadia National Park (of which this study area is a part). Make the "Mainelab" the active theme.

13. Peruse the database. Refer to the data dictionary (appendix A) for details.

14. Now, what the guides want is a map that shows medium to tall, deciduous, broadleaf trees that occur as either continuous or interrupted densities. Referring to the data dictionary, decide which code numbers to which you need to refer to make the map suggested in #12 above. Now, use your query builder to produce a query of the database that satisfies these criteria.

15. We can do the same with the herbaceous vegetation as well. Let us say, for example, that an ecologist specializing in clump grasses is trying to develop a search or sampling strategy. What the ecologist needs is a grouping of all the grasses based on their heights and based on their densities. Because the ecologist is not interested in sod-forming grasses, we do not need to search for continuous or interrupted densities, only scattered and rare. The ecologist would like

to see the following outlined on the map so that an appropriate stratified sampling strategy can be developed for the clump grasses.

1. tall scattered grasses
2. tall rare grasses
3. medium scattered grasses
4. medium rare grasses
5. short scattered grasses
6. short rare grasses

16. Use the query builder to create this map.

17. Until now we have been performing reclassifications based on the attributes of each individual polygon. Now we are going to reclassify the database using criteria about the neighborhoods around features. This type of reclassification is not as common as the previous method, but it does provide some very different insights into your database.

18. Make the "Mainelab" theme active.

19. In the SE Desert Island Maine database there is a substantial lake in the north-central portion of the map called Eagle Lake. See if you can find this lake (it is long and oriented north and south). Select this polygon using the select feature icon and look at its table contents. Note its polygon number so you can refer to it later.

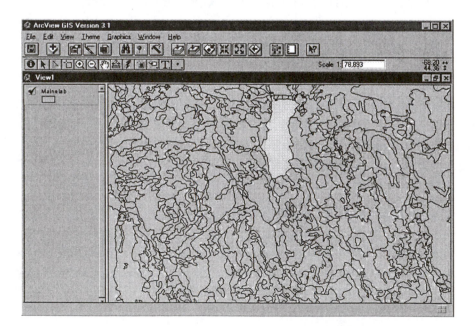

20. Lakes often provide good habitat for both animals and humans (particularly the vacationing kind of humans), because they have plenty of woody shade from the trees that often grow nearby. If might be nice to see what the polygons neighboring the lake are like and what they contain. We could, of course do this by simply going in and selecting each individual polygon that is visually adjacent to the lake. However, you can easily use the "Select by Theme" dialog box from the Theme menu to do this for you.

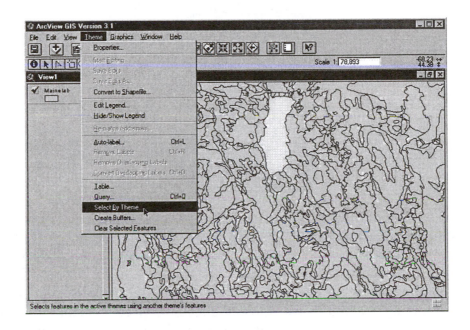

21. When you call up the dialog box you will select "Within Distance Of" from the first dropdown menu. Because there is only one theme in the view the "Selected Features Of " dialog will be preset to "Maine." What you want to do is to guarantee that the distance between the selected polygon (the lake) and the adjacent polygons is zero (adjacent). So you use a selection distance of 0 in the dialog menu.

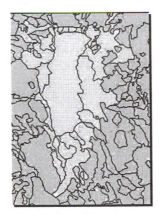

22. Once you have completed this operation you should see a set of polygons surrounding the lake, together with the lake itself, highlighted. Because you don't want the lake, you want only

the adjacent polygons, you need to deselect the lake itself from the selected polygons. You can do this by using the query builder by creating an expression that says Mainelab# <> ___ and fill in the appropriate polygon number for your lake. **Be sure to say "select from set" when you do this**. Now you should have a ring of polygons (a spatial neighborhood) surrounding the lake. You can easily imagine how a combination of this plus the classification work you have also done in this laboratory could do for a realtor trying to define the best places to buy lake-front for development of resort areas.

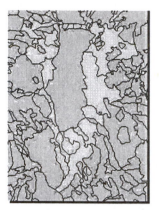

23. This ring of polygons surrounding the lake is a rough approximation of a buffer, but it is based solely on adjacency. We could modify the query to go out further distances, but the way ArcView works under this approach is simply to select the polygons that are within this specified distance. It does not create new polygons based on the actual buffer distance. There are many other ways of building neighborhoods. Most of these involve combinations of thematic and distance measures. Time prohibits a detailed run through all of these many options for this laboratory. We'll consider them another time. Still, there are two more useful approachs you should know. First, after selecting Eagle Lake, use the buffer wizard (under the themes pull-down menu) to create a 0.5 mile buffer around Eagle Lake. Make this buffer an outside buffer, around the selected feature of the theme "mainlab," and save it in a separate view. When you are finished the default version of the output masks the lake itself. To see the lake in the background double click on the legend icon for the buffer theme and make it an outline rather than a solid color. When you are done your buffer should look something like the one below. This is the normal approach to creating a buffer. When you are finished, take a look and practice with some of the other options for buffering using the buffer wizard.

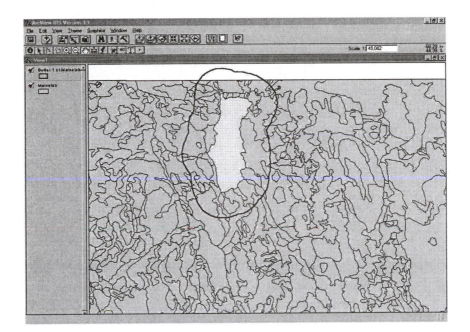

24. There is another method of creating buffers. This is an optional exercise that you may choose to perform to give you more practice. Suppose you put in a large parking lot using the drawing tools available in ArcView. Draw in a parking lot (any shape you want, any reasonable size, and anywhere you would like to place it).

25. Now use the "Select Features Using Shape" button to select the polygons that are inside your parking lot.

26. By opening the theme table to display the theme's attribute data, what can you tell about large trees that had to be felled to produce your parking lot?

27. Shut down and close out your work.

Products:

Complete and turn in the discussion questions below, together with completed themes.

Review and Discussion Questions:

1. Suggest ways that you might be able to query the Mt. Desert Island database to answer questions like...

    Is there a difference between vegetation on the coast and inland?

    Is there a particular pattern to barren areas in the database?

    How could we determine appropriate dear habitat?

2. Give an example of how you could use the data included in the database to find a more appropriate parking lot than the random one you produced?

3. Many creatures like a diversity of habitat, rather than just a single habitat. Describe how you could use the drawing tools and the select features to give you a measure of habitat diversity (defined by the number of different land cover types) around a point location where a deer was spotted.

4. Using a sample of the data from this database create a table that conforms to first normal form. Now discuss the problems associated with creating that same table using the entire database.

5. Describe how reclassification based on neighborhoods differs from reclassification based on the polygonal attributes themselves.

Exercise 8, Appendix A:

Southeastern Mount Desert Island, Maine Abbreviated Data Dictionary

Polygons:

Polygon Numbers: (MAINE#) Polygons' sequential input value

Polygon ID: (Maine-ID) #'s indicating Küchler's Field Notes as applied to each polygon identified by aerial photography.

In this regard there are often several polygons associated with each Main-ID number because these were used for a-posteriori aerial photo interpretation whereby he first identified the similarity of polygons prior to visiting them and performing a detailed vegetation survey.

Species Code Description: (up to 29 species per polygon) {S1...S29)

7 digit code

First three digits indicate genus (missing or uncertain genus names start with a code number of 99 followed by numeric value indicating the alphabetic listing of the item) (e.g. grasses = 991).

Next two digits indicate species (and/or subspecies). (missing or uncertain species names have a code number of 99).

The sixth digit is the species coverage code (see below)

The seventh digit is the species sociability code (see below)

<u>Species Coverage Code Definition:</u>

1 = Under 1% of the area
2 = 1-5% of the area
3 = 6-25% of the area
4 = 26-50% of the area
5 = 51-75% of the area
6 = Over 75% of the area

<u>Species Sociability Code Definition:</u>

1 = Species scattered singly
2 = In small bunches or tufts
3 = In patches
4 = In larger groups
5 = In extensive sheets

<u>Woody Vegetation Code Description: (Up to 7 per polygon) {W1...W7)</u>

4 Digit Code:

1st digit: seasonality

1 = Evergreen
2 = Deciduous

2nd digit: leaf form

1 = Broodleaf
2 = Needleef

3rd digit: height code

1 = Tall (minimum height of trees is 25 meters)
2 = Medium (median height of trees is 10-25 meters)
3 = Low (maximum height of trees is 2 meters)
4 = Shrub (minimum height of shrubs is 1 meter)
5 = Dwarf Shrub (maximum height of shrub is 1 meter)

4th digit: density code

    1 = Continuous
    2 = Interrupted
    3 = Scattered singly
    4 = Rare

Herbaceous Code Description: (Up to 7 per polygon) {H1...H7)

3 Digit Code:

1st Digit: leaf form:

    1 = Graminoids
    2 = Forbs
    3 = Lichens and Mosses
    4 = Barren
    5 = Aquatic Vegetation

2nd Digit: Height Code

    1 = Tall (minimum height of herbaceous plants is 2 meters)
    2 = Medium (height of herbaceous plants is 0.5 to 2 meters)
    3 = Low (maximum height of herbaceous plants is 0.5 meters)

3rd Digit: Density Code

    1 = Continuous
    2 = Interrupted
    3 = Scattered Singly
    4 = Rare

Exercise 8, Appendix B:

Southeastern Mount Desert Island, Maine Abbreviated Data Dictionary: Detailed Species Codes

SPECIES CODING OF POLYGONS

****

SPECIES LIST FOR SE MOUNT DESERT ISLAND, MAINE

TOTAL SPECIES = 242 (EXCLUDING 8 UNKNOWN SPECIES)

TOTAL GENERA = 144 (EXCLUDING 4 UNKNOWN GENERA)

PLANTS WITH UNKNOWN GENUS OR SPECIES

(CODE=99199) **Floating Fillamentous Algae**

(CODE=99299) **Grass spp**

(CODE=99399) **Lichen spp**

(CODE=99499) **Moss spp**

****

PLANTS WITH KNOWN GENUS BUT UNKNOWN SPECIES

(CODE=02699) *Carex* **spp**

(CODE=03399) *Cladonia* spp

(CODE=03999) *Crataegus* spp

(CODE=12199) *Scirpus* spp

(CODE=12299) *Solidago* spp

(CODE=13999) *Mnium* spp

(CODE=14099) *Panicum* spp

(CODE=14199) *Poa* spp

(CODE=14299) *Polytrichum* spp

(CODE=14399) *Sphagnum* spp

(CODE=14499) *Triticum* spp

****

(CODE=00101) *Abies balsamea*

(CODE=00201) *Acer pennsylvanicum*
(CODE=00202) *Acer rubrum*
(CODE=00203) *Acer saccharum*
(CODE=00204) *Acer spicatum*

(CODE=00301) *Achillea millefolium*

(CODE=00401) *Agropyron repens*

(CODE=00501) *Agrostis alba*
(CODE=00502) *Agrostis scabra*
(CODE=00503) *Agrostis tenuis*

(CODE=00601) *Alnus crispa var mollis*
(CODE=00602) *Alnus rugosa var americana*

(CODE=00701) *Ambrosia artemisiifolia var elatoir*

(CODE=00801) *Amelanchier laevis*

(CODE=00901) *Ammophila breviligulata*

(CODE=01001) *Anaphalis margaritacae var intercedens*

(CODE=01101) *Andromeda glaucophylla*

(CODE=01201) *Antennaria neodioica*

(CODE=01301) *Anthoxanthum odoratum*

(CODE=01401) *Apocynum androsaemifolium*

(CODE=01501) *Aralia hispida*
(CODE=01502) *Aralia nudicaulis*

(CODE=01601) *Arctostaphylos uva-ursi var coactilis*

(CODE=01701) *Arenaria groenlandica*

(CODE=01801) *Arisaema strorubens*

(CODE=01901) *Aster acuminatus*
(CODE=01902) *Aster lateriflorus*
(CODE=01903) *Aster macrophyllus*
(CODE=01904) *Aster nemoralis*
(CODE=01905) *Aster novi-belgii*
(CODE=01906) *Aster novi-belgii var litoreus*
(CODE=01907) *Aster radula*
(CODE=01908) *Aster umbellatus*

(CODE=02001) *Athyrium filix-femina var michauxii*
(CODE=02002) *Athyrium thelypteriodes*

(CODE=02101) *Betula lutea*
(CODE=02102) *Betula papyrifera*
(CODE=02103) *Betula populifolia*

(CODE=02201) *Brachelytrum erectum var septentrionale*

(CODE=02301) *Brasenia schreberi*

(CODE=02401) *Calamogrostis canadensis*

(CODE=02501) *Calopogon pulchellus*

(CODE=02601) *Carex cephalantha*
(CODE=02602) *Carex conoidea*
(CODE=02603) *Carex crinita*
(CODE=02604) *Carex gracillima*
(CODE=02605) *Carex lenticularis*
(CODE=02606) *Carex lurida*
(CODE=02607) *Carex paupercula var irrigua*
(CODE=02608) *Carex pennsylvanica*
(CODE=02609) *Carex projecta*
(CODE=02610) *Carex rostrata*

(CODE=02611) *Carex rostrata var utriculata*
(CODE=02612) *Carex scorparia*
(CODE=02613) *Carex stricta*
(CODE=02614) *Carex trisperma*
(CODE=02615) *Carex vulpinoidea*

(CODE=02701) *Celastrus scandens*

(CODE=02801) *Ceraslium vulgatum*

(CODE=02901) *Chamaedaphne calyculata var angustifolia*

(CODE=03001) *Chrysanthemum leucanthemum var pinnatifidum*

(CODE=03101) *Cirsium arvense*

(CODE=03201) *Cladium mariscoides*

(CODE=03401) *Clintonia borealis*

(CODE=03501) *Comptonia peregrina*

(CODE=03601) *Coptis groenlandica*

(CODE=03701) *Cornus canadensis*

(CODE=03801) *Corylus cornuta*

(CODE=04001) *Dactylis glomerata*

(CODE=04101) *Danthonia spicata*

(CODE=04201) *Dennstaedtia punctilobula*

(CODE=04301) *Deschampsia flexuosa*

(CODE=04401) *Diervilla lonicera*

(CODE=04501) *Drosera intermedia*
(CODE=04502) *Drosera rotundifolia*

(CODE=04601) *Dryopteris disjuncta*
(CODE=04602) *Dryopteris spinulosa*
(CODE=04603) *Dryopteris noveboracensis*
(CODE=04604) *Dryopteris phegopteris*

(CODE=04605) *Dryopteris therlipteris var pubescens*

(CODE=04701) *Dulichium arundinaceum*

(CODE=04801) *Eleocharis palustris*
(CODE=04802) *Eleocharis smallii*

(CODE=04901) *Empetrum nigrum*

(CODE=05001) *Epilobum angustifolium*

(CODE=05101) *Equisetum sylvaticum forma multiramosum*

(CODE=05201) *Erigeron canadensis*

(CODE=05301) *Eriophorum spissum*
(CODE=05302) *Eriophorum virginicum*

(CODE=05401) *Eupatorium perfoliatum*

(CODE=05501) *Fagus grandifolia*

(CODE=05601) *Festuca capillata*
(CODE=05602) *Festuca rubra*

(CODE=05701) *Fraxinus americana*

(CODE=05801) *Galmia angustifolia*

(CODE=05901) *Gaultheria hispidula*
(CODE=05902) *Gaultheria procumbens*

(CODE=06001) *Gaylussacia baccata*
(CODE=06002) *Gaylussacia dumosa var bigeloviana*

(CODE=06101) *Glyceria striata*
(CODE=06102) *Glyceria projecta*

(CODE=06201) *Hammelis virginiana*

(CODE=06301) *Hemoracallis fulva*

(CODE=06401) *Hieracium aurantiacum*
(CODE=06402) *Hieracium floribundum*

(CODE=06403) *Hieracium pilosella*
(CODE=06404) *Hieracium pratense*

(CODE=06501) *Hierochloe odorata*

(CODE=06601) *Hudsonia ericoides*

(CODE=06701) *Hypericum gentianoides*
(CODE=06702) *Hypericum perforatum*
(CODE=06803) *Hypericum virginianum var fraseri*

(CODE=06901) *Ilex verticillata*

(CODE=07001) *Impatiens capensis*

(CODE=07101) *Juniperus communis var depressa*
(CODE=07102) *Juniperus horizontalis*
(CODE=07103) *Juniperus mucronata*

(CODE=07201) *Juncus balticus var littoralis*
(CODE=07202) *Juncus canadensis*
(CODE=07203) *Juncus effusus var solutus*
(CODE=07204) *Juncus gerardi*
(CODE=07205) *Juncus tenuis*

(CODE=07301) *Kalmia angustifolia*

(CODE=07401) *Larix laricina*

(CODE=07501) *Lathyrus japonicus var glaber*
(CODE=07502) *Lathyrus palustris*

(CODE=07601) *Lechea intermedia var juniperina*

(CODE=07701) *Ledum groenlandicum*

(CODE=07801) *Linnaea borealis var americana*

(CODE=07901) *Lupinus polyphyllus*

(CODE=08001) *Luzula multiflora*

(CODE=08101) *Lysimachia quadrifolia*
(CODE=08102) *Lysimachia terrestris*

(CODE=08201) *Maianthemum canadense*

(CODE=08301) *Mitchella repens*

(CODE=08401) *Myrica gale*
(CODE=08402) *Myrica pennsylvanica*

(CODE=08503) *Najas flexilis*

(CODE=08601) *Nemopanthus mucronata*

(CODE=08701) *Nuphar variegatum*

(CODE=08801) *Nymphaea odorata*

(CODE=08901) *Onoclea sensibilis*

(CODE=09001) *Oryzopsis asperifolia*

(CODE=09101) *Osmunda cinnamomea*
(CODE=09102) *Osmunda claytoniana*
(CODE=09103) *Osmunda regalis var spectabilis*

(CODE=09201) *Phleum pratense*

(CODE=09301) *Picea glauca*
(CODE=09302) *Picea mariana*
(CODE=09303) *Picea rubens*

(CODE=09401) *Pogonia ophioglossoides*

(CODE=09501) *Pinus banksiana*
(CODE=09502) *Pinus resinosa*
(CODE=09503) *Pinus rigida*
(CODE=09504) *Pinus strobus*

(CODE=09601) *Plantago juncoides var decipiens*

(CODE=09701) *Poa compressa*
(CODE=09702) *Poa palustris*
(CODE=09703) *Poa pratensis*

(CODE=09801) *Polygonum cilinode*

(CODE=09901) *Polypodium virginianum*

(CODE=10001) *Polystichum acrostichoides*

(CODE=10101) *Pontederia cordata*

(CODE=10201) *Populus tremuloides*
(CODE=10202) *Populus grandidentata*

(CODE=10301) *Potentilla simplex var calvescens*
(CODE=10302) *Potentilla tridentata*

(CODE=10401) *Prunus pennsylvanica*
(CODE=10402) *Prunus virginiana*

(CODE=10501) *Pteridium aquilinum var latiusculum*

(CODE=10601) *Pyrus americana*
(CODE=10602) *Pyrus communis*
(CODE=10603) *Pyrus malus*
(CODE=10604) *Pyrus melanocarpa*

(CODE=10701) *Quercus rubra var borealis*
(CODE=10702) *Quercus rubrum*

(CODE=10801) *Rannunculus acris*
(CODE=10802) *Rannunculus bulbosus*

(CODE=10901) *Raphanus raphanistrum*

(CODE=11001) *Rhododendron canadense*

(CODE=11101) *Rhinanthus crista-gallis*

(CODE=11201) *Rhus typhina*

(CODE=11301) *Rhynchospora alba*

(CODE=11401) *Rosa carolina*
(CODE=11402) *Rosa nitida*
(CODE=11403) *Rosa palustris*

(CODE=11501) *Rubus allegheniensis*
(CODE=11502) *Rubus hispida var obovalis*
(CODE=11503) *Rubus hispidus*
(CODE=11504) *Rubus idaeus var strigosus*

(CODE=11505) *Rubus pubescens*

(CODE=11601) *Rumex acetosella*

(CODE=11701) *Sarracenia purpurea*

(CODE=11801) *Salix alba var vitellina*
(CODE=11802) *Salix discolor*
(CODE=11803) *Salix gracilis*

(CODE=11901) *Sambucus canadensis*
(CODE=11902) *Sambucus pubens*

(CODE=12001) *Sarracenia purpurea*

(CODE=12101) *Scirpus cespitosus var callosus*
(CODE=12102) *Scirpus cyperinus var pelius*
(CODE=12103) *Scirpus validus var creber*

(CODE=12201) *Solidago bicolor*
(CODE=12202) *Solidago canadensis*
(CODE=12203) *Solidago graminifolia*
(CODE=12204) *Solidago juncea*
(CODE=12205) *Solidago nemorialis*
(CODE=12206) *Solidago randii*
(CODE=12207) *Solidago rugosa*
(CODE=12208) *Solidago semper virens*
(CODE=12209) *Solidago squarrosa*
(CODE=12210) *Solidago uliginosa*
(CODE=12211) *Solidago uliginosa var linoides*

(CODE=12301) *Smilacina trifolia*

(CODE=12401) *Sparganium chlorocarpum*
(CODE=12402) *Sparganium fluctuans*

(CODE=12501) *Spiraea latifolia*
(CODE=12502) *Spiraea tomentosa*

(CODE=12601) *Stellaria graminae*

(CODE=12701) *Thatictrum polygamum*

(CODE=12801) *Thuja occidentalis*

(CODE=12901) *Thymus serpyllum*

(CODE=13001) *Trientalis borealis*

(CODE=13101) *Trifolium agrarium*
(CODE=13102) *Trifolium hybridum*
(CODE=13103) *Trifolium pratense*
(CODE=13104) *Trifolium repens*

(CODE=13201) *Triglochin maritima*

(CODE=13201) *Tsuga canadensis*

(CODE=13301) *Typha latifolia*

(CODE=13401) *Utricularia purpurea*

(CODE=13501) *Vaccinium angustifolium*
(CODE=13502) *Vaccinium corymbosum*
(CODE=13503) *Vaccinium oxycoccos*
(CODE=13504) *Vaccinium vitas-idaea var minus*

(CODE=13601) *Viburnum cassinoides*

(CODE=13701) *Vicia cracca*
(CODE=13702) *Vicia tetrasperma*

(CODE=13801) *Viola cucullata*
(CODE=13802) *Viola incognita*
(CODE=13803) *Viola pallens*

Laboratory Exercise 9
Spatial Arrangement

Introduction:

In laboratory exercise 8 we saw how we could manipulate classification to our benefit. By changing the classification to suit our particular needs we were able to produce a unique set of attributes for each task. This is a powerful capability of the GIS, but there are still more to come. In laboratory exercise 9 we will examine some of the capabilities of ArcView to help us define spatial arrangement. Because the map is the most powerful tool available for examining spatial relationships in the analog domain, we should be able to examine some of the spatial characteristics available within a single coverage inside an automated GIS.

There are an increasing number of spatial arrangement metrics being applied to distributions of phenomena on the surface of the earth. These metrics most often provide comparative mathematical descriptions of the numbers, proportions, densities, interspersions, and the like. While there are many metrics, few of them have been implemented directly into the GIS, but some are available through manipulations of macro languages (such as "Avenue" for ArcView), or as separate programs linked to the analytical capabilities of the GIS itself. We will limit ourselves in this laboratory to those techniques that are within the readily available analytical methods of ArcView.

For this laboratory we will be using a dataset from the Northern Virginia Military District of Central Ohio. Details of work that has been done on this databases can be obtained from the following sources (Boerner, et al. 1995, Simpson et al. 1994, and DeMers et al. 1996), all listed in your text. The study area is divided into two distinct regions: a northern half (labeled region 1 in your database) that is north of Darby Creek on undulating glacial till, and a southern half (labeled region 2 in your database) south of Darby Creek on a very flat glacial outwash plain. What we are going to do in this exercise is to examine some of the internal compositional elements of each of these two regions to determine how they differ both structurally and functionally. Because we also have coverages for two different time periods (1940 and 1988) we will also be able to determine how the northern and southern halves changed through time, giving us the ability to see if the physical setting will change how the two halves respond.

Learning Objectives:

Using a portion of the Northern Virginia Military District database we will examine the functional capabilities of ArcView for performing analysis of internal spatial patterns. We will tabulate a number of structural characteristics for analysis, and we will use sampling techniques to examine the structure of the map a piece at a time.

When you are finished with this laboratory you should be able to:

1. Determine the percentages, density, and diversity of landcover types.

2. Compare the percentages of landcover on different landscape types.

3. Compare the changes in landcover from one time period to another.

4. Compare the amounts of linear objects (fencerows in this case) from landscape to landscape.

5. Compare the changes in amount of fencerows through time.

6. Develop alternative spatial arrangements that can be evaluated within this database.

Linkage to your text:

While the capabilities of ArcView are somewhat limited, many of the 17 learning objectives from chapter 11 in your text are reinforced here. For example you will clarify your understanding of arrangments (#1), describe different types of distributional patterns (#'s 2, & 3) and you will answer questions regarding line pattern analysis (#'s 9, 10, 11, 12 and 13). Text learning objectives 14 through 17 are best studied when we examine the Network Analyst in Lab exercise 13.

Methods:

Analyzing Polygonal Arrangements:

Comparison of amounts:

1. In today's lab you will be working with an ARC/INFO database named Ohio that has been exported to ArcView. This database has been copied for you and is found in your course directory. It contains polygonal information about land cover and/or land use, and linear

information about fencerows. These were recorded from 1940 and 1988 aerial photography for a small poriton of the Northern Virginia Military District in central Ohio.

2. Open ArcView and open a new view. This database was input in the UTM coordinate system with map units and distance units both in meters. Go to the View Properties pulldown menu and enter these values. Add the following themes to your view: Luse40, Luse88, Fence40, and Fence88. Convert all of these themes to shapefiles (ArcView themes) and save these files to your workspace on your own computer. Delete the original (ARC/INFO coverages) themes from your own workspace. As you may recall from Lab 8, this is done to prevent changes to the original databases.

3. Make the landuse 1940 theme (Luse40) the active theme. Wander through the tables for the landuse 1940 theme. Notice the landuse codes available for each polygon. The following are the meanings of the landuse codes. Technically these are landcover values rather than landuse and we will use that terminology from now on. Note: if you want to view the different landcover types double click on the legend for each active theme and change the legend from "single symbol" to "unique value," using the Luse-code for the values field.

201 Upland Forest
202 Young Woodland
203 Oak Savanna Park
204 Riparian Woodland
205 Riparian Pasture
206 Agriculture
207 Urban
208 Early Successional
209 Borrow Pit
210 Industrial

How many landcover types are found? _____

4. The database is designed so that you can separate out the northern half (the till plain) from the southern half (the outwash plain). If you examine the tables you will see that there is a code indicating the number 1 for the northern half and the number 2 for the southern half. See the figure above for example.

5. In the table below fill in the types of landuses, their total acreage, and the % of the database each comprises for both the northern and southern portions of the study area. You will need to separate out the northern half of the study area from the southern half using your query tool.

LANDCOVER 1940

LANDCOVER		HECTARES		% COMPOSITION	
till plain	Outwash plain	till plain	Outwash plain	till plain	outwash plain

6. Now make the 1988 landcover theme active. Wander through its tables. How many landcover types are found? _____. What accounts for the differences between 1940 and 1988?

7. As before, make a table (or use ArcView) indicating landcover, acreage and % composition using the blank table below.

LANDCOVER 1988

LANDCOVER		HECTARES		% COMPOSITION	
till plain	outwash plain	till plain	Outwash plain	till plain	outwash plain

Comparison of densities and diversities:

1. As you can observe from the table above there is a considerable difference in the overall composition of landcover types on the northern versus the southern halves of the study area (i.e. based on geomorphological position), as well as based on the changes from 1940 to 1988. Among the major changes in land use you can easily observe are primarily those of agriculture. In both years and in both portions of the study area agriculture dominates. This domination has an immense impact on the relative and absolute spacing of other, smaller, land use polygons. Your text indicates a number of techniques that would be useful for examining the arrangement of points (e.g. nearest neighbor, quadrat analysis, etc.), lines (e.g. connectivity, line intercepts,) and polygons (e.g. contiguity using join count statistics). While many of these techniques could be performed with a fair amount of preparatory work, we'll restrict our arrangement analysis to things that can be performed without writing macros. We'll begin by estimating the densities of area objects in space. A first measure requires us mearly to analyze the average size of each polygon based on its type (i.e. what is the average size of riparian vegetation polygons?). Another simply measures the diversity of landuse types within a specified distance.

2. We will look at both of these. With the landcover theme(s) active, go to the theme tables and separate out region 1 from region 2 for each of 1940 and 1988. You can use the table query tool to select by zone.

3. Now, from the table, highlight the field "Area" and choose "statistics" from the "Field" menu. This will allow you to find the minimum, maximum, and average size polygon for each of the two regions for each of the two time periods (1940 and 1988). Note what you find so you have it for later reference.

4. To examine the diversity of landscape units (landcover types in our case) we need to sample our polygons with a standard size object. For this we will use the drawing tools (illustrated by a dot on the second tier of icons). If you click on the dot and hold for a second or so, a pull down set of icons shows you alternative shapes to draw. Draw a circle of radius approximately 1,780 meters (this will give you a polygon of approximately 10 square kilometers). Rather than spending several hours trying to get the circle exactly the size you want you can use an easier method. First, create any circle and place it randomly on the theme. Now select the graphic and click the right mouse button. Use the pulldown menu to select the properties of this circle. From there you can type in the actual radius of the circle you want and place it precisely where you want. Once you have your circle, place it anywhere on the north half of the map of 1988 landcover.

5. Now randomly place 4 more of these on the north half of the map of 1988 landcover.

6. Each of these circles represents approximately 10 square kilometers of area and what we want to know is, on average, how many different landcover types are there per 10 square kilometers of area. Select the polygons using the "select with graphic" icon.

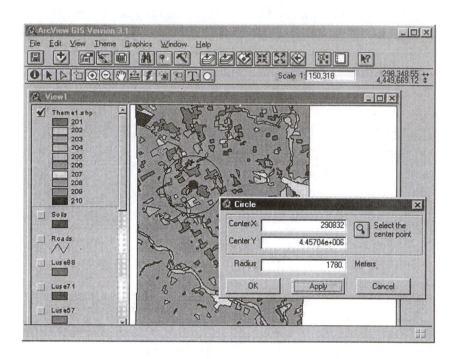

7. Calculate and record the average number of unique landcover types of the 10 samples you performed. You may want to create a table to catalog these.

8. Repeat steps 3 through 7 for the southern half of the study area.

9. What you have calculated is a simple measure of landscape diversity (called the gamma diversity by landscape ecologists). What is the difference in average gamma diversity between north and south?

Analyzing Linear Arrangements:

1. Now we are going to calculate some simple metrics for linear objects. In this case we have a set of fencerows (fences in which vegetation has grown). We have these, as before, for 1940 and 1988. Be sure these themes are in your database as available themes.

2. For the 1940 theme, for each portion of the database (till plain and outwash plain), go to the theme tables and determine the amount of fencerow (in meters) and calculate it as a percentage of the total fencerow for the entire study area. Complete the table below (or use the ArcView software to make your own).

FENCEROWS 1940

METERS		% OF TOTAL	
till plain	outwash plain	till plain	Outwash plain

3. Repeat this process in step 2 for the 1988 theme.

FENCEROWS 1988

METERS		% OF TOTAL	
till plain	outwash plain	till plain	Outwash plain

4. Note the differences between the 1940 and 1988 databases. Think about how you might be able to calculate other metrics like nearest neighbor distances between line, or calculate resultant vectors of line fields. Are these possible within ArcView?

5. Before you shut down, consider the following review and discussion questions. You may need to have your databases open to answer some of them.

Products:

Provide answers to the following questions as well as any digital thematic maps you instructor requires of you.

Review and Discussion Questions:

1. What did the polygonal metrics (i.e. the tabulated data and the gamma index) show you about the composition of the northern and southern ½ of your databases and about the differences between the years 1940 and 1988?

2. What did the analysis of linear data tell you about the differences between the northern and southern ½ of the database as well as the temporal differences?

3. What relationships did you see between the areal patterns you observed and the linear patterns?

4. Suggest some alternative methods for analysis of this database (e.g. what would you like to be able to do to answer your questions)? **HINT: these make very good project topics.**

5. Provide extensions (a wish list if you will) of techniques for each of these exercises that were not readily available to you with ArcView that you would really like to see.

6. Finally, provide at least two extensions of the techniques used in each of these two laboratory exercises that could be done with database manipulation, rather than with macro commands. (e.g. how could you change the NVMD database to create neighborhoods of fencerows?).

Laboratory Exercise 10
Overlay

Introduction:

In laboratory exercise 9 we examined some elementary capabilities of ArcView to help us define spatial arrangement. These techniques allow us to examine the spatial arrangements of objects within a single coverage (although we can compare the tabulated data for each coverage after the fact). It is necessary, however to be able to compare variables between and among coverages or themes as well. This, in fact, was the primary motivation for the original development of the Canada GIS in the 1960's.

The primary method by which data from one coverage are compared to data from another is through the use of a variety of "Overlay" functions. Overlay functions come in a number of different types (see text) and can be applied in both the raster and vector models. For this laboratory we are going to use a "logical" type of overlay called "intersection." Intersection will find all thematic polygon data that are common to both themes. It is the equivalent of the logical "and" statement in set theory. The opposite of intersection is union overlay that finds all thematic polygon data that are included in both themes. Union overlay is the equivalent of the logical "or" statement in set theory. As you use the overlay wizard inside ArcView, take the time to read what the software says about what it is doing. The software also includes a very nice set of graphics that will help you understand what is happening.

For this laboratory, as in laboratory 9, we will be using the Northern Virginia Military District dataset from Central Ohio. Remember that details of work that has been done on this databases can be obtained from the following sources (Boerner, et al. 1995, Simpson et al. 1994, and DeMers et al. 1996), all listed in Fundamentals of Geographic Information Systems. The study area is divided into two distinct regions: a northern half (labeled region 1 in your database) that is north of Darby Creek on undulating glacial till, and a southern half (labeled region 2 in your database) south of Darby Creek on a very flat glacial outwash plain. What we are going to do in this exercise is to examine the relationships between the 1940 landuse/landcover data and the 1988 landuse/landcover data by performing an intersection overlay operation on them. This is a direct approach to examining how land use has changed, rather than looking at the internal configuration of each, tabulating it and comparing the tables, as we did in laboratory 9.

Learning Objectives:

Using a portion of the Northern Virginia Military District database we will examine the functional capabilities of Arc/View for performing logical overlay analysis, particularly as it relates to the intersection method commonly used in ARC/INFO. We will create a new theme by overlay (intersecting) the 1988 landuse coverage on the 1940 landuse coverage to see what changes have taken place from 1940 to 1988. **You MUST save your final themes because one or more of them will be used in the next laboratory exercise on cartographic output.**

When you are finished with this laboratory you should be able to:

1. Create new themes by selecting specific portions of existing themes and save them as shape files.

2. Perform vector overlay on these new themes using the geoprocessing wizard.

3. Explain the results of the overlay operations in terms of land cover dynamics.

4. Explain the difference between union and intersection overlay operations.

Linkage to your text:

While we will focus primarily on only two types of polygon overlay operation to keep the software learning curve to a minimum, you should also examine other forms of overlay such as point-in-poly and line-in-poly and others, especially because you may want to use them for your project. You should also familiarize yourself with the theoretical material associated with all overlay types in your text. In general, then this exercise is linked to all the objectives in chapter 12 of your text.

Methods:

Using the geoprocessing wizard:

      Preliminary setup:

1. In today's lab you will be working with the ARC/INFO coverages from lab 9. This database has been copied for you and is found in the lab9 directory, and contains polygonal information about landcover and/or land use, and linear information about fencerows and edges between distinct landcover types. These were recorded from 1940 through 1988 aerial

photography for a small portion of the Northern Virginia Military District in central Ohio. We will be using only a few of these themes, but others are available (in various stages of completion) should you wish to experiment.

2. Open ArcView and open a new view.

3. Add the following themes to your view: Luse40 (land cover 1940) and Luse88 (land cover 1988).

4. Convert (if they are not already) all of these themes to shapefiles (ArcView themes) and save these files to your workspace within your course directory.

5. Delete the original (ARC/INFO coverages) themes from your view. (As you will recall from Laboratory 8, this is done to prevent changes to the original databases.)

6. In number 3 above you loaded multiple themes. This step is necessary to allow the use of certain extension options (i.e. many overlay operations) for step number 3 below.

7. Under the file pulldown menu go to extensions and load the geoprocessing extension.

8. Under the view pulldown menu go to the geoprocessing wizard and highlight the "intersection" procedure.

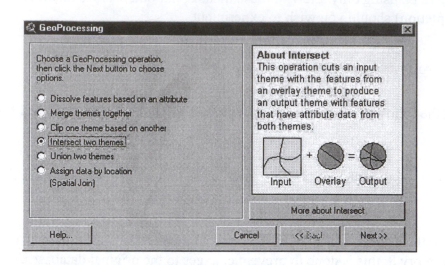

9. Before proceeding read what the procedure is for and how it works. You may also want to look at the other options such as "union" and think about using some of them for your project.

To make sure you understand what the overlay procedures are doing (and because some of the polygons seem to exceed the software's limit for number of vertices, resulting in some really strange results) we are not going to just haphazardly overlay all thematic data all at once. Instead, we are going to look at specific thematic information from one time period to the next. In this way we will be using the overlay operation to help us gain a better understanding of how land cover has changed through time in our study area. Additionally, we will be separating our study area into two distinct areas (north and south) so we can see how these different landscape units respond differently through time. This will give us more information than we had in laboratory 9 where we simply tabulated the changes in land cover. Instead of just providing gross numbers of land cover change we will now be able to examine what land covers changes to what other land covers. This will provide us with a small level of predictability.

Preparing our themes for overlay.

1. Make luse40 (the 1940 land cover theme) the active theme in your database.

2. Now use the query builder to select region 1 from the active theme. You will notice that the top half of the theme becomes yellow indicating that all the polygons on the northern half of the database have been selected.

3. Save the selected polygons as "NLC40" standing for north land cover for 1940 by saving it as a shapefile (when the software asks you if you want the theme included in your database say yes. Now you have a theme that shows just the land cover for 1940 for the undulating till plain north of Darby Creek.

4. Repeat steps 10 through 12 for luse88, labeling your new shapefile "NLC88" standing for north land cover for 1988.

5. Now repeat the same process as for steps 10 through 13 for the southern have of the two themes producing two new shapefiles to include in your database. These will be called "SLC40" and "SLC88" respectively for land cover themes on the outwash plain for 1940 and 1988.

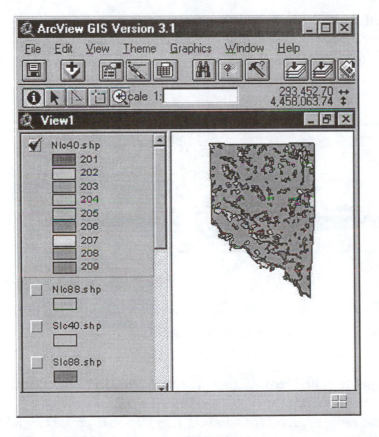

Before going any further, if you have not labeled these themes by land cover type from the previous laboratory you should **do so now** so you know what land covers you have rather than just the codes. We are going to examine specific changes in land cover over time. While the software will allow you to just overlay all cover types simultaneously it does run into problems with very large polygons and also, you as a GIS analyst are going to have to decide what changed. The process we are going to use is a binary approach, allowing you to have full control over each individual thematic map comparison. While this may not seem as elegant, it is much easier for you (and of course a client) to understand.

The specific questions we want to answer are as follows:

- How much has urban expansion encroached on agriculture from 1940 to 1988 in the till plain (north)?

- How much has urban expansion encroached on agriculture from 1940 to 1988 in the outwash plain (south)?

- Has there been any change in the amount of forest from 1940 to 1988? How much? Where is this change most notable (outwash plain or till plain)? Specifically we want to look at the change from young woodland (202) to upland forest (201).

1. Make the north 1940 theme (NLC40) active.

2. Use the query tool to isolate all polygons within the NLC40 theme that represent agriculture (code 206).

3. Save as a shapefile called NAG40 (agriculture in 1940 in the north) and place it in your view.

4. Now, using the same approach as numbers 1 and 2 above create two additional shapefiles: urban for 1940 in the north (NURBAN40) and urban for 1988 (NURBAN88).

5. Repeat the process for the southern (outwash plain) and create the following shapefiles for agriculture in the south for 1940 (SAG40), urban for 1940 (SURBAN40) and for 1988 (SURBAN88).

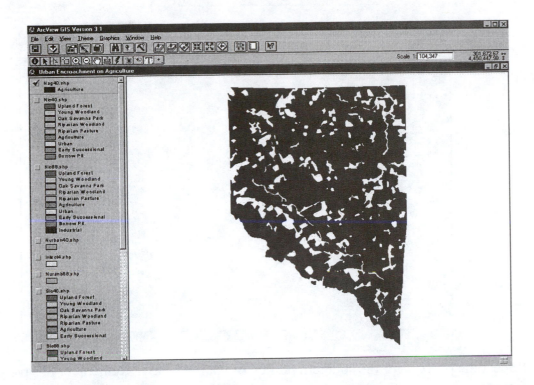

6. To examine the change in urban (urban expansion) switch between the 1940 and 1988 urban themes for both the north and the south just so you can visualize this change.

7. Now use the GeoProcessing wizard (under the view pulldown menu) to perform an intersection overlay using the 1940 agriculture (north) as the input them, with the 1988 urban (north) as the overlay theme to produce an output shapefile called NAGURB (north ag-urban).

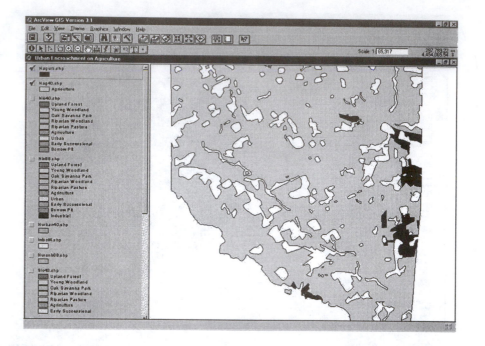

8. Repeat the process for the southern portion of your database. Remember to save your themes.

9. Now, repeating the general operations you performed for the relationship between 1940's agriculture and 1988's urban, examine how much young woodland (202) was allowed to proceed to maturity and become upland forest (201). Save your themes.

10. After finishing these overlay operations examine the tables for these output themes and proceed to the review and discussion below.

Products:

Hand in the answers to the questions below. The questions are meant not only to help you understand the functionality of cartographic overlay operations, but also to give you some insights about when and why you would use the techniques in the first place. Spend some time thinking about the questions rather than just answering them blindly.

Review and Discussion Questions:

1. Describe what the process of logical overlay using intersection created for both the land use example and the forest succession example.

2. What does this tell us about the process of urban encroachment on agricultural land on the northern portion of this study area from 1940 to 1988? What about the southern half?

3. Are there any differences from north to south? What accounts for them? How could examine these?

4. Describe how you could examine other land cover types that are being converted to urban.

5. Are there any differences in patterns of vegetation succession from north to south between 1940 and 1988?

6. We would normally expect to see young woodland continue to change to upland forest. Where does it not? How can cartographic overlay be used to examine these situations?

7. Why did you not use "union" overlay for these two scenarios? What would it have told you? Describe a situation where union overlay would be useful.

8. How could you perform a line-in-polygon intersection? What might you expect to see? Describe how the fencerows could be overlaid with the with selected land cover polygons to provide answers to specific questions. Be precise in your answer as to exactly what questions you are trying to answer.

Laboratory Exercise 11
Output From Analysis

Introduction:

In laboratory exercise 10 we performed a logical overlay using intersection on two themes to produce a final coverage. While this and many of the other analytical techniques available to us in GIS are exciting they are not an end unto themselves. The final product from analysis (usually to a second party user) is most often a cartographic document. While we may understand the analysis and even the original output from analysis it is important to communicate this to the user who may not be as familiar with it as we. For this reason we strive to produce a coherent, easy to understand, and graphically telling map. While this is not a course in cartography it is important as the final product of your work. If you have not had a course in cartography you may want to read chapter 12 carefully before proceeding.

Learning Objectives:

Using the results of polygon overlay from exercise 10 on a portion of the Northern Virginia Military District database we will now create a final map layout. The layout will include all the basic map elements normally found in a cartographic document and will show you the strong linkage between cartography and GIS.

When you are finished with this laboratory you should be able to:

1. Create layouts for the output of your analysis.

2. Manipulate the cartographic elements in your layouts.

3. Add titles and other annotations to your layouts.

4. Include multiple themes in the same layout.

Linkage to your text:

This exercise links implicitly with chapter 12 of your text. While the theoretical portion of the text is not explicitly called on in the exercise it is not a bad idea to refer to it as you develop your cartographic layout and final map product.

Methods:

Creating a layout:

We are going to create a layout (map) of the results of one of our previous pieces of analysis. In this case we are going to produce a layout of the encroachment of urban into agriculture in our study area (the northern portion of the Northern Virginia Military District). We want to show the change areas that were in agriculture in 1940 that changed to urban by 1988. In addition we want to include the following:

- A locator map on the upper right hand side. The locator map should also include:
  1. Both the northern and southern portions of the study area.
  2. A neat line around the map.
  3. A grid system.
- A modern north arrow.
- An appropriate title.
- A scale bar in kilometers.
- A legend.
- A neat line around the entire map.

In other words, our final map output should look something like this:

1. Open ArcView and open the project you worked on in the last laboratory and create a new view. Add the following themes to your view: Nagurb and Nag40. In the properties menu rename this view LAB 11. You will change this name later on.

2. Open a new View and add the Luse88 theme. This view is going to be used to create the locator portion of your layout. From the View properties menu rename this view LOCATOR MAP so you remember what it is for. You will also need to change the properties of this view to meters for map units and meters for distance units. Also, within this view keep the legend a single symbol because you don't want to see all the land cover types.

3. In the LOCATOR view use the query tool to select the northern half of the Luse88 theme. This is the portion that will be displayed as output from your analysis from lab 10. Make the theme active and display it.

4. Now return to the LAB 11 View. Make Nag40 and Nagurb your active themes and zoom to these active theme(s). If your themes legends are not labeled be sure to label Nag40 as "1940 Agriculture" and Nagurb as "Urban from Ag" so these will show up on your layout.

5. From the View menu, choose Layout. A template manager appears to help you decide how you want your map layout to look.

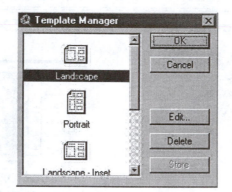

6. From the template manager select the landscape format at the very top.

7. Unless you have already created a layout before, a menu should indicate new layout as your only option. Select this option. Immediately a layout of your map will appear. This layout is linked to the view that you are working on. This means that if you go back to the original view and change something, it will also be changed on the layout. Try the following to see what happens.

8. Go into your view and select properties from the view pulldown menu. Once there change the name of the view from LAB 11 to Urban Encroachment on Agriculture (1940 – 1988). Now go back to your layout and use the layout you already created (layout1) rather than a new layout. Notice how the title of your map changed.

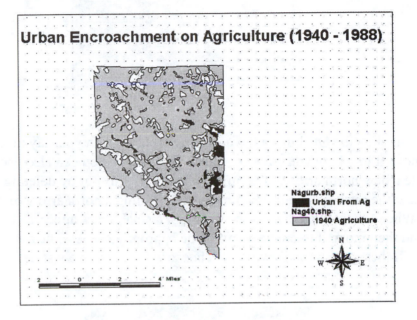

9. Notice how the title tends to be a little big to fit on the map properly. You need to resize this. Using your select tool select the title and then use the resize bars to make the title a bit smaller and to move it around on the screen.

10. You also want to change the north arrow to something that is a little more modern. With the pointer tool active double click on the existing north arrow. When you do this a selection of north arrows appears. This allows you to select a different north arrow such as the one shown below.

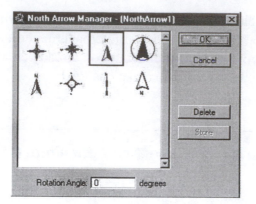

11. When you click the OK button the north arrow changes on your layout to the one you want. Now move it to where you want on your layout. Note: you can also use the pulldown menus to create a new north arrow. You might want to try this out by deleting the existing north arrow and then using the pulldown menu and creating a north arrow of whatever size you wish. When you activate the north arrow tool a small cross appears. Insert this wherever you wish and then move it to create a box of the appropriate size. The north arrow selection menu will then appear to allow you to choose your north arrow style.

12. You should also notice that the legend is in miles. We want this to be in Kilometers. Double click on the scale bar and when the scale frame bar menu appears change the settings to Kilometers and the number of intervals to 2. After clicking the OK button resize and move the scale bar where you want.

13. Your map is starting to come along, but you want to include a locator map now. To make a little room move the map object and the legend where you like. This is where the earlier preparations to create the separate view comes in.

14. Within your layout you should have some empty space on the upper right hand corner. Using the view frame pulldown menu draw a rectangle just a little smaller than this space. You want to leave some room for some other stuff we will add. Once you create the rectangle the View Frame Properties menu shows up. Select the LOCATOR MAP view you created earlier in the laboratory.

15. A small version of the map shows up in the rectangular space you created. We want the locator map to tell us more about the study area. For example it would be really nice if the map had a grid system on it. It turns out ArcView has a tool you can use just for that. Before we go on, however let's see if we can save this layout so we can find it again when we need

it. You have probably already seen that your layout goes away if you leave the layout mode. Of course you can keep your layout if you just minimize the layout rather than exiting from it. But if you like the layout you have and might want to use it again you can store your layout as a template called Lab 11. You do this from the layout pulldown menu.

16. Now, that you feel safe get into the layout mode (if you are not already there). Under the file pulldown menu load the graticules extension.

17. When you do this the following icon appears on your menu bar.

18. Now click on the icon. When the menu appears select the measured grid and the LOCATOR MAP options. This means that we are going to create a measured grid around the LOCATOR map, not the other map in our layout.

19. Press the Next button to view the options for your measured grid. Change the grid interval to 6000 so that there will be less numbers for this small graphic. Then press Preview if you want to see what it is going to look like before you finish.

20. Other options appear on the next menu. Just accept the default settings for the time being. You can try the options later if you wish.

21. You now have a small locator map that also has a measurement grid displayed around it. Now you want to be sure to separate the locator map from the rest of the layout. Do this by first selecting the locator map with a single click. Then go to the menu bar and click on the neat line icon.

22. After selecting this icon choose the options you wish for style of neat line around the locator map.

23. Finally, use the neat line icon to create a neat line around your entire map by choosing the "place around all graphics" option. Manipulate the outside neat line until it resembles what you want. You can save this as before by making it a template with whatever name you want.

Exporting

As you might guess, not everyone (especially your clients) have access to the GIS software you use. You often have to prepare a report that includes the maps as part of the document. While you can print your layouts directly from ArcView it is often desirable to simply embed them directly into your document. This is what you will do next.

1. Begin by selecting the file pulldown menu and the export option.

2. You will notice that the menu allows you to decide what you will call your file, where you will put it and what type of file you will create. Your instructor may have some specific ideas as to what format is most desirable for them. Try saving the file in each of the formats available to you. You will probably need to keep these somewhere on your hard disk because some formats are quite large.

3. When you have done this examine each file type you created to see how big it is. There is often a direct relationship between the file size and the output quality.

4. Save the file in the format your laboratory instructor wishes you to use.

Products:

The map you produce here is meant to augment what you discovered in laboratory exercise 10. Although there are some cartographic design ideas that can be explored here, the primary task is that you produced an exportable map that can now be input to a document for review by a client. In this case your client is your laboratory instructor. For your output create a 5-10 page word-processed document that includes the final layout you exported. The document should describe what you did in laboratory 10, and what the results indicate. If you really want to impress your instructor you might go back and think of how impressive a table showing changes in agricultural land might be. The details of output will be up to your laboratory instructor. You are also required to answer the questions below as part of your graded product.

Review and Discussion Questions:

1. Why did you have to create a separate view for your locator map?

2. Describe how you might include charts, graphics, or even photographs in your output.

3. Consider some alternatives to the output suggested in this exercise that might be more informative for the user.

4. Why did you have to put a neat line around the locator map?

Laboratory Exercise 12
Working With Surfaces: (ArcView® Spatial Analyst)

Introduction:

Until now we have pretty much ignored surfaces in our laboratory exercises. The reasons for this are practical rather than conceptual. Within ArcView itself there really isn't a way to analyze surfaces, nor raster data of any kind. For that you need a special extension to ArcView called Spatial Analyst. Spatial Analyst is like a subset of its larger cousin ArcGRID and both are relatively similar to Dana Tomlin's original Map Analysis Package (MAP), although with a different user interface and often different approaches. While Spatial Analyst has a great many tools for analyzing grid or raster data we will be using it primarily so we can examine some of the concepts and ideas about statistical surfaces—primarily topographic in this case. Once you learn the basics of Spatial Analyst you might want to continue your education by working through the tutorial exercises provided with the software.

Learning Objectives:

Using Spatial Analyst and some data provided with it and ArcView by ESRI we will examine some of the basic operations available to us for operating on surfaces within the raster data model. We will re-examine the raster tesselation model itself and will convert existing grid data to shapefiles to see what effect such conversion has on the data. We will create contour maps from grid data, add hill shading for visual effect, slice our surface into large intervals and use the Map Calculator to examine the Map Algebra language for modeling.

When you are finished with this laboratory you should be able to:

1. Describe how Spatial Analyst stores both rational and integer grid data.

2. Convert a grid dataset into shapefiles.

3. Create a contour map of surface data.

4. Create a shaded relief map of surface data.

5. Slice a topographic surface into selected units.

6. Create maps of topographic slope categories (neighborhoods).

7. Create maps of topographic aspect categories (neighborhoods).

8. Reclassify slope and aspect neighborhoods into binary maps.

9. Use the Map Calculator to combine binary slope and aspect categories.

Linkage to your text:

This laboratory incorporates all of the basic ideas of chapter 10 in Fundamentals of Geographic Information Systems, not all of the learning objectives are employed. We will focus on the concept of the statistical surface and its representation (learning objectives 1-4), surface slicing (learning objective 8). Volume calculations, dasymetic maps, dot distribution maps and other forms of surface representation are not explicitly covered in this laboratory, nor are the many forms of interpolation or viewshed analysis. In addition to learning objectives of chapter 10 we will be revisiting cartographic overlay—specifically mathematical overlay from chapter 12 (specifically learning objective 2). Finally, we will prepare for our discussion of cartographic modeling from chapter 13 by creating a very simple portion of a cartographic model using the Map Calculator.

Methods:

Loading Spatial Analyst and Examining its contents.

1. Load ArcView with a new view. When asked if you want to add data say no.

2. Before adding data you want to have the Spatial Analyst extension to ArcView added. As before, go to the file pulldown menu, navigate to extensions and check the Spatial Analyst box to load the extension.

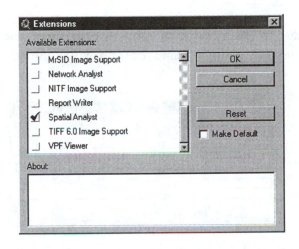

3. Now you can add themes that are based on the raster or grid data structure. Click on the add themes button or use the view pulldown menu to add new themes. Under the add theme dialog box select the Grid Data Source as the data source type and navigate to the avtutor/spatial subdirectory on your server or PC.

4. Select the elevgrd theme and display it. It should look something like this.

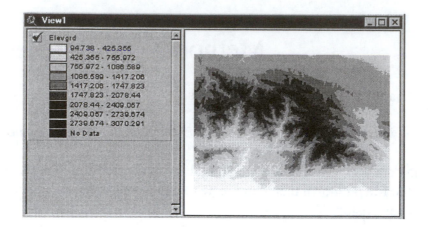

5. Use the identify button to examine the contents by clicking on various places on the map. You should notice that only a single value for each (x,y) coordinate pair appears. You should

also notice that the value is a rational number (decimal fraction) and also that the tables icon is inactive. This is because Spatial Analyst stores these numbers as unique values. This saves space when decimal numbers are being used. Later on we will look at integers to see how they are stored in Spatial Analyst.

6. While the elevation map you see on the screen may be satisfactory it is not what we are normally accustomed to seeing for a contour map. Notice how the menu bar now has two new buttons (analysis and surface). We are going to use the surface features first. Making sure your theme is selected select the create contours option from the surface pulldown. When the dialog box appears change the contour interval from 100 to 250 so you can see some of the detail of your contours. Leave the base contour as 0.

7. Now deselect Elevgrd and make the Contours of elevgrd the active theme. Display it alone.

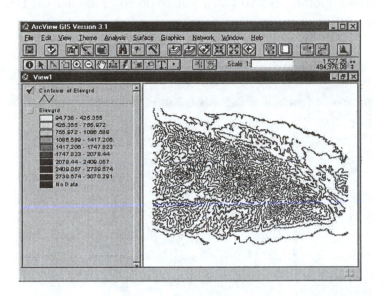

8. As an option you can check Elevgrd and show it in the background. This may help you visualize your surface data.

9. Another way to help you visualize topographic data is to add hillshading. Use the Compute Hillshading option under the Surface pulldown menu. Accept the defaults (you can always try other options later). Display the results.

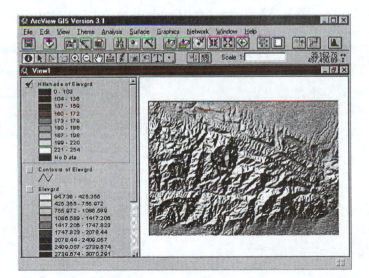

10. The Surface pulldown menu also includes two neighborhood classification functions specifically designed for use with surface features—Derive Slope and Derive Aspect. In many models you will develop it is necessary to isolate specific amounts and aspects of topographic slope. Say, for example that you want to find regions with steep, north facing

slopes only. These operations will allow you to do so. First use the Derive Slope command and display the results.

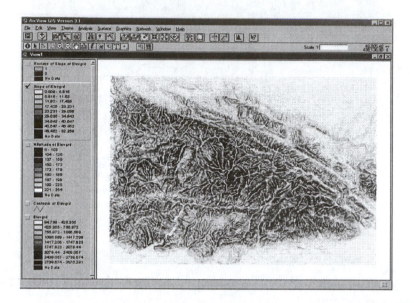

11. The slope command selects predefined slope classes (neighborhoods) based on an equal interval classification method (the default). You can change this default by double clicking on the legend to get the legend editor. We are going to choose a different approach because we want to select between two separate types of slopes (steep and gentle). For this we will use the Classify function under the Analysis pulldown menu. Change the new values for slopes between 0.009 and 23.231 degrees to 0 and all the rest except missing values to 1.

12. You should notice that you now have a binary map that looks something like the one below. For the sake of clarity you should relabel the map (double click on the legend) by labeling gentle for 0 and steep for 1. Remember we are looking for steep slopes. You have changed the values in your map from rational numbers to integers. Notice for example that the tables icon is now active. Go ahead and click on it to see how your data look. You could now add attributes to this table if you wanted to. Think about how this extends the simple raster model by allowing the grid cells to have multiple attributes.

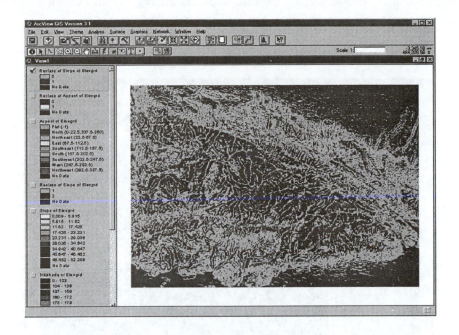

13. Now that you have created a binary map of slope from the Elevgrd thematic map it is now time to create one for aspect. Make the Elevgrd theme active and use the Derive Aspect option. Display the map.

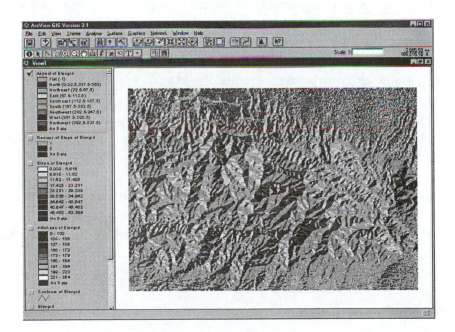

14. As before we are going to use the Classify function under the Analysis pulldown menu. As you remember we want to find north facing slopes only. So let's assign 1 to north and 0 to

all other aspects. And as before double click on the legend and change the labels to
BadAspect for 0 values and GoodAspect for 1 values.

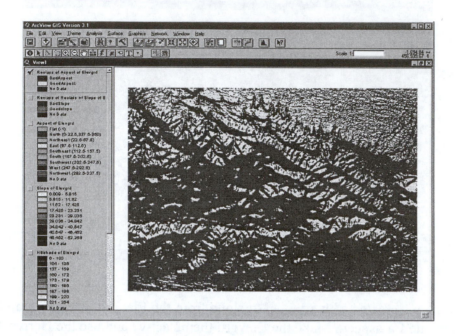

As you may have guessed, there was a good reason to select 0 and 1 for the classified data. With
1's indicating good slopes (meaning steep slopes) as well as good aspects (north facing slopes –
actually +/- 40 degrees). We should be able to combine these values to find both good slopes
and good aspects. In vector we could have performed an intersection overlay operation. Here we
can use the mathematical overlay operation by multiplying these two coverages together. This is
done by using a new and powerful tool called the Map Calculator found under the Analysis
pulldown menu. This tool is based on Dana Tomlin and Joseph Berry's Map Algebra map
analysis language. We'll just finish our little raster model by using the Map Calculator to
multiply the two themes you just created into a final thematic map showing the best slope /
aspect combination.

15. Under the Analysis pulldown menu select the Map Calculator.

16. Expand the Map Calculator so that the entire text for all the available themes shows up.

17. Double click on (Reclass of Slope of Elevgrd).

18. Click the multiply button.

19. Double click on (Reclass of Aspect of Elevgrd).

20. Click the Evaluate button on the bottom of the Map Calculator. Hint: to make multiple expressions you would want to take special care to place your cursor outside the existing part of the equation to make sure it is evaluated first.

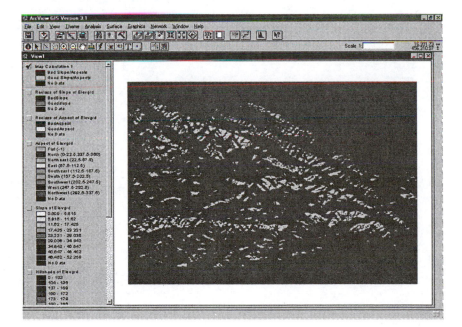

21. The results of your evaluation should show the mathematical equivalent expression (in Map Algebra) of a logical intersection. In this case only good slopes (1's) and good aspects (also 1's) will show up as 1's. All the bad slopes and aspects will literally zero out. This produces the following map that shows good slope/aspects (1's) and bad slope/aspects (0's). You will notice that I have labeled this. You should do the same both for practice and so you remember what you just created.

22. Now select an area and zoom into it four or five times. Notice the grid cells.

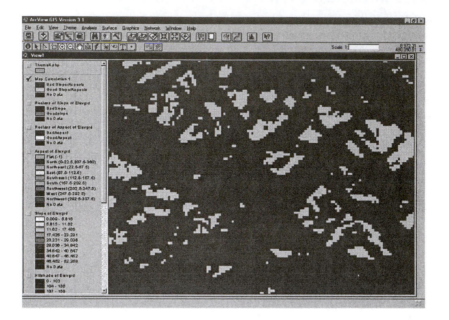

23. Now use the Theme pulldown menu and convert the raster theme into a shapefile. Notice the difference.

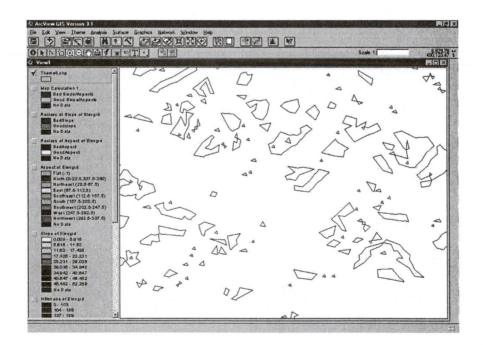

24. Proceed to the questions.

Products:

Hand in the answers to the questions below. The questions are meant not only to help you understand the functionality of cartographic overlay operations, but also to give you some insights about when and why you would use the techniques in the first place. Spend some time thinking about the questions rather than just answering them blindly.

Review and Discussion Questions:

1. Describe and provide a diagram of how integer data are stored in Spatial Analyst. Now do the same for rational data.

2. Describe the difference between vector cartographic overlay using the geoprocessing wizard and raster cartographic overlay using the Map Calculator.

3. When you perform the contour of your surface, why does the software give you the option of changing the base elevation value? What possible situations would suggest something other than 0?

4. When you computed hillshade what impact does changing the azimuth have on the hillshade effect? What impact does changing the altitude have?

5. Describe the difference between raster and shapefile representations of the final output of your analysis in terms of:

   - Shapes and sizes of polygons especially along the edges.
   - Storage of attribute data as tables.
   - The effects of cartographic overlay.

6. From chapter 13 examine the process of model flowcharting. Now provide a simple flowchart for the model you developed in steps 10 through 23.

7. Why did you create binary maps to create your model?

8. List and describe some possible scenarios where binary maps could be created to compare non-topographic themes to derive an outcome.

Laboratory Exercise 13
Working With Networks (ArcView® Network Analyst)

Introduction:

As with surfaces we have left networks alone, and for the same practical reason. ArcView itself doesn't deal with route finding, allocation of service areas along a network, or evaluating accessibility along a network. For those functions you need a special extension to ArcView called Network Analyst. Network Analyst is like a subset of the NETWORK module of ARC/INFO. While there are specific needs for setting up network databases to solve problems within this extension we will not be examining these explicitly in this laboratory. Instead we will evaluate some of the conceptual issues of working with networks. I recommend reading and practicing the Network Analyst tutorial provided with the software to enhance your abilities with this software.

Your textbook, Fundamentals of Geographic Information Systems, indicates that networks are not just collections of linear objects, but rather higher level line objects. The higher level derives from the linkages between individual arcs, from the characteristics of the nodes that connect these arcs, and from the attributes of the linear objects themselves. Example networks include, but are not limited to, railways, roads, streams, pipelines, hedgerows, and electrical utility lines. All of these allow the movement of vehicles, animals, people, electricity, and even ideas from one place to another. They are also embedded within a background matrix of land uses, homes, plants and animals and, therefore interact with these landscape features. Examples might be the accessibility to homes through networks of roads, the movement of small mammals along hedgerow networks and between fields lined by these networks, the dispersion of particulate pollutants along roadside and rail verges. The laboratory exercise you are about to complete will give you experience using Network Analyst in particular, but, more importantly, networks in general. Try to remember that while learning this software, like all the others available to you is designed to help you understand the geography of networks so that, as the software improves, you will be prepared to employ it in the solution of these types of problems.

Learning Objectives:

Using Network Analyst and some data provided with it and ArcView by ESRI we will examine some of the basic operations available to us for operating on higher order linear objects called networks. We will examine the concepts of connectivity and accessibility, optimum routing and allocation. Although we will not be developing network databases we will examine some of the necessary database characteristics for them so that you can extend your modeling capabilities beyond the simple exercises you will employ here.

When you are finished with this laboratory you should be able to:

1. Locate "to" and "from" locations within a network database by…

    *   pointing on the screen
    *   entering addresses
    *   using point information

2. Provide driving directions for a truck.

3. Find the "best" path from one place to another using the following options…

    *   shortest path
    *   most interesting path
    *   fastest path

4. Create a service (customer) base near a site.

5. Determine how many customers there are within a service (customer) abase.

6. Suggest different types of functional distance along a network.

Data Preparations:

Besides the typical criteria for functional distance (e.g. time and distance) you can use other criteria if you wish by specifying an appropriate cost field in a line theme's feature table as long as the field has some measurable cost units.

In addition to allowing different ways of defining best path, Network Analyst allows you additional options. For example, if your need is to provide flower delivery to a set of clients and you have agreed on a date and time of day, you can set up a visit schedule or sequence of visited locations. In this approach you minimize the time between each pair of visits. A wholesale grocer, unlike our florist, is more interested in getting as many groceries delivered to as many locations as possible in a day. In this case Network Analyst allows you to visit locations through the most efficient order.

Before you can do any of this, however, you have to decide where the sites are and you must specify them explicitly. Depending on the type of data you may have there are several ways to do this:

- If your locations are in shapefile or ARC/INFO formats you can add point themes to your view using the Add Theme button or the View pulldown menu.

- If your locations are in dBASE format or are stored as a delimited text file (with coordinate fields), you can add your point theme by using the Add Event Theme option.

- If the files mentioned above have specific address information, you can geocode the locations and add the geocoded theme as before but by using the Address option from within the Add Event Theme.

- Finally, if you don't have a set of point data stored as a file (again see the Northern Virginia Military District database as an example) you can use the Network Analyst Add Location button ▣ or the Add Location Address button ▣ to create the point data one at a time. This approach is also used for specifying specific locations within a point theme to visit and even the sequence in which you will visit them.

Linkage to your text:

This laboratory incorporates the ideas of point-in-polygon and line-in-polygon from Chapter 12, linear measurement from Chapter 8. The vast majority of the laboratory includes the concepts of linear arrangement, connectivity, routing, and allocation from Chapter 11.

Methods:

Loading Network Analyst and preparing for analysis.

1. Load ArcView.

2. From the File menu select Extensions and load the Network Extension by checking its box. (Hint: if you want to have any of the extensions added automatically check the "make default" check box on the right. This will save you from having to do this every time.)

3. Navigate to the avtutor/network subdirectory in your server or PC.

4. Once there you will notice a project highlighted on the left called qstart. This has the extension .apr which means it was saved as an ArcView project file for you to work on. Select this project file.

5. When you have done this you will notice that 4 views become available to you. Each is designed for specific exercise.

You don't need special network data for Network Analyst. Any ArcView line theme (e.g. the roads theme in the Northern Virginia Military District Database you have been working with), or any linear ARC/INFO coverage. To make things easy we are going to use the network theme data provided with the Network Analyst package. Click on the add themes button or use the view pulldown menu to add new themes.

Giving Directions Based on Shortest Route.

Some of you may have used software provided by your computer vendor when you purchased it or available in web sites that provides you with routing information from one place to another. In case you didn't notice it, you were using a GIS. We are going to do the same thing here with Network Analyst. In our case we are going to act in the role of a dispatcher in San Francisco to assist one of your drivers caught in rush-hour traffic reach his destination. He is looking for the shortest route.

1. From within the qstart.apr project select the "Truck Route" view. A network database (streets theme) with the San Francisco, California shoreline background appears.

To provide a set of directions you must first...

- Locate your driver (start point) and where the driver is going (destination). You will do this first by using the cursor.

- Of course if you are going to provide directions you must also find the shortest route from the start point to the destination.

- Then provide the directions.

2. Make the streets theme active.

Our driver is westbound on the Bay Bridge. You need to specify this location. First, how does the software know where any of these locations are? With the streets theme active open up the theme's table. You don't need to know all of the variables, but you should notice that addresses and distance measurements are included explicitly.

3. Under the Network pulldown menu chose the Find Best Route option. This opens a problem definition dialog box and adds a result theme called Route 1 to the view. The final solution of your shortest path route will be stored here.

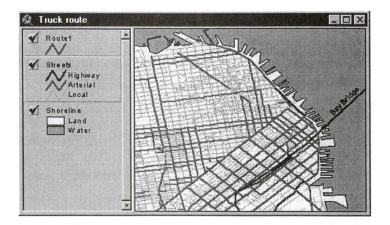

4. Now click on the Add Location tool 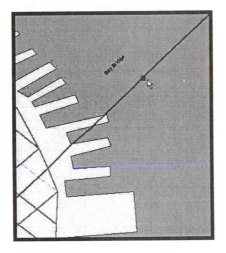, then move the cursor to the truck (over the San Francisco Bay on the Bay Bridge.

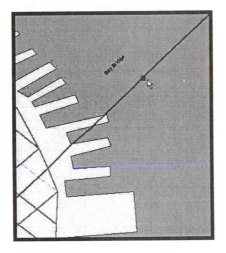

5. The driver must deliver the package to 1862 Polk St.. But the theme doesn't actually show you where this location is. Fortunately, the streets table includes this information. The easiest way to find this location is to use the Add Location by Address button , and entering the address, then pressing OK.

Locate Address ☒

Enter address:

| 1862 Polk St. |

Preferences... — OK — Cancel

You have just performed a process called **geocoding** in which you converted an address into a point theme and added it to a view.

6. Now that you have selected the start location and the destination you need to find the shortest route. You do this by clicking the Solve button . Notice below how the program shows the length of the route as almost 4.2 km. In addition, the program also provides the final route as a map theme (remember that route 1 had earlier been set up to save this information).

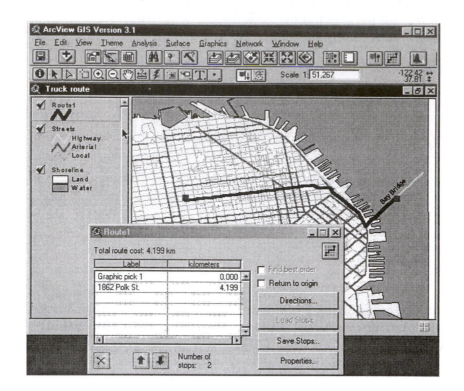

7. However, our dispatcher needs to provide this information to the truck driver as a set of directions. To do this, simply click the Directions button. This provides the travel directions.

8. Notice how the Directions menu also allows you to print the directions and to save them as a separate text file. Your laboratory instructor may want you to try this out so you can include it in a laboratory report.

Creating an efficient delivery route.

You are delivering flowers to a number of customers. Because of unusually high volume on the north side of the city you want to assign a driver to that regional alone. To do this you need a delivery plan that minimizes the driving time. This amounts to providing the most efficient order of stops in the shortest route possible.

To set up the delivery plan you must first…

- Locate your truck depot (start point) by entering its address.

- Use a point theme that includes your customers. These will be your delivery stops.

- Find the most efficient order of delivery for your flowers and the shortest route to deliver them.

1. Open the Delivery route view.

2. Make the Streets theme active.

3. Using the Network pulldown menu select Find Best Route. As before a problem definition dialog appears. And as before ArcView adds a theme called Route1 to the view that will eventually contain the route for your driver. (Remember to be sure your streets theme is active or this will not work).

4. To specify your starting point (the depot) click the Add Location by Address button , enter 60 Spear St. and click OK.

5. Now you need to load stops so press the Load Stops button in the problem definition dialog box and select the Deliveries theme. Remember that this is the list of stops that you have to visit.

6. Unlike our previous exercise, in this case the truck needs to return to the depot when they are done with their deliveries. Check the Return to origin checkbox so that your last destination is the depot.

7. Now check the Find Best Order checkbox to find the most efficient route to deliver your flowers.

8. Click the Solve button to determine the shortest route and report its length.

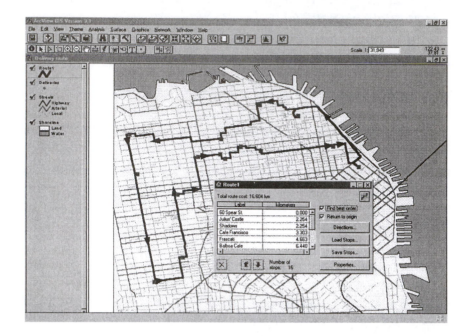

9. As before you can print or save the result. Check with your laboratory instructor to see if you will be required to save this. (Note: It is important to keep in mind that such routing algorithms as the one you have just used are not exact algorithms. In fact, the traveling salesman problem is very complex and requires simplifications and shortcuts in their solution, especially where there are many stops).

Getting the nearest ambulance to an accident.

The scenario is fairly common. There has been an injury automobile accident during rush hour. Your task as the ambulance dispatcher is to find the closest ambulance service, and to get that ambulance to the scene as quickly as possible. In this case the shortest path in distance may not be the shortest in terms of time. Time is critical in this case.

To find the closest ambulance service and get it to the scene as quickly as possible you must...

- Use the point theme to show the available hospitals.

- Enter the address of the accident.

- Determine drive times using the street theme's attribute tables.

- Find the nearest hospital.

- Calculate the quickest route to the accident scene.

1. From the qstart project open the Closest Hospital view.

2. Make the Streets theme active.

3. Choose the Find Closest Facility from the Network menu. ArcView adds a theme called Fac1 to the view that will eventually contain the ambulance route.

4. Now check Travel to event (accident). This tells ArcView that the travel will be from the hospital to the scene of the accident.

5. The accident was at Polk & Lombard. Use the Add Location by Address button to enter this information. Given that this is not an address, how does ArcView figure the location out? Look at the tables for the streets theme to see what you can determine from that.

6. Press the Properties button in the problem definition dialog box to specify DRIVETIME for the cost field. Notice how the working units changes.

7. Now press the Solve button to determine the closest ambulance and the quickest route to the scene of the accident.

Creating a service area and determining how many customers live there.

As a real estate consultant you are tasked with determining how accessible a potential retail property is to its surroundings. Specifically you want to determine two things: (1) 5 and 10 minute drive time zones (functional distances) and (2) how many people live in those same zones. This will allow the real estate firm to evaluate the viability of this property.

To find the answers to these two questions you will…

- Enter the address of the site.

- Create the drive time service areas.

- Determine the number of customers within the service areas.

1. Begin by opening the Site Accessibility view from the qstart project.

2. Make Streets active.

3. Choose the Find Service Area option from the Network pulldown menu. In this case two themes are added to the view: (1) Snet1 that will contain the streets within the selected drivetime and (2) Subarea1 that will contain the drive time polygons.

4. The potential site is located at 672 Sutter St. Use the Add Location Address button to enter this address.

5. Double click in the Minutes field (not on the title "minutes), delete the default settings, and type in 5, 10. These are the drive times in minutes within which you are to determine your customer base. Then press the solve button.

6. Turn off the Snet theme by removing the check from its check box. Drag the Customers theme to a position just above the Subarea1 theme. Notice how this changes the display.

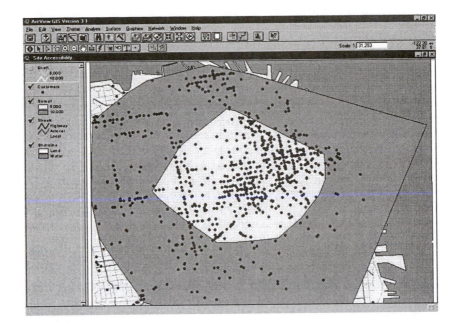

7. Now we need to determine how many customers there are in these service areas. Begin by clicking on the Customers theme and make it active.

8. From the Theme pulldown menu choose Select By Theme. A Select By Theme dialog box appears. Pick the Subarea1 theme.

9. Choose Are Completely Within and press the New Set option. This selects all customers (points) within the drive time polygons. This is a point-in-polygon analysis that you might remember from chapter 10 in Fundamentals of Geographic Information Systems.

10. Click the Table button and you will see the number of customers that fall within the 10 minute driving time.

11. Now make Subarea1 active.

12. Click the Select Feature tool and use your cursor to select the 5 minute driving time polygon only.

13. Repeat steps 7, 8, and 9 to select the number of customers that fall within the 5 minute driving time polygon from the site.

14. Finally, click the Table icon and the number of customers within the 5 minute driving time polygon will be displayed in the Number selected box.

15. Proceed to the questions.

Products:

Hand in the answers to the questions below. The questions are meant not only measure what you remember about what you just did, but to demonstrate your deeper understanding of the concepts of networks and how they might affect the results of your network analysis.

Review and Discussion Questions:

1. What units were used as cost or friction values for each of the following?

   - shortest path
   - fastest path
   - most interesting path

2. Provide two additional examples of path types that could be used and explain what the measurable cost units would be.

3. Describe the difference between finding the a timed route with multiple stops and finding the most efficient route with multiple stops.

4. Given the routes you have just manipulated what impact do you think increased connectivity would have on the results of your results? Provide a simple diagram illustrating how this might work.

5. Describe how you believe ArcView calculates service areas. Provide a diagram illustrating your point.

6. Describe how ArcView calculates how many customers there are within a customer base.

Laboratory Exercise 14
The Cartographic Modeling Process

Introduction:

The cartographic modeling process operates in two separate, but inextricably linked processes—**model formulation** and **model implementation**. These processes generally operate in opposite directions and are most easily visualized through the use of flowcharts. The exercise you are about to perform will require you to both formulate and implement a simple linear land use planning model. In addition, it will require you to develop flowcharts for both the formulation and the implementation. While cartographic modeling can be performed in both raster and vector, it is often easiest to conceptualize in raster because it was originally created for a raster environment, and because the model created in raster can frequently be more complex than those created in vector. Because this will be your first attempt at creating a relatively complete cartographic model we will use the more straightforward raster approach. The grid database you are about to use is rather coarse in resolution, but has sufficient themes to perform much of what is necessary for a rather complete land use evaluation and site assessment model.

While the model, as with all models, requires the use of **descriptive** components, its use is to make specific decisions about the use of the land. Your purpose is to create a ranked set of site assessments for potential non-agricultural land uses. By default, the highest ranked areas will be suggestive of the best possible locations for non-agricultural land uses. In that case then the model becomes **prescriptive** in that it prescribes the best locations. What follows is a brief description of the model you are about to develop and implement.

The Land Evaluation and Site Assessment (LESA) model is a real-world, although no longer used, land use and evaluation model designed to provide a set of planning alternatives for communities concerned about urban expansion into high quality agricultural lands. As populations increase, more and more of the highest quality lands will be consumed by this urban expansion. The United States Department of Agriculture (USDA), Soil Conservation Service (SCS), now called the Natural Resources Conservation Service (NRCS), in direct response to Federal Legislation developed the LESA model as a simple, additive, linear model to evaluate both the physical attributes of the land (Land Evaluation) as well as the socio-economic factors that may have a part in planning the land to prevent massive loss of the most productive lands. The LESA model then, was at one time a congressionally <u>mandated</u> model. The NRCS was <u>required</u> at that time to implement this model to determine potential federal agricultural subsidies. (For further information on the legislation see the 1983 and 1985 farm bills).

The primary problem with the LESA system, besides its rather simplistic approach to landscape management, is that it requires the SCS to perform an analysis of each parcel of land individually (as well as manually). Lee Williams (1985 see citation in Fundamentals of Geographic

Information Systems), then at the University of Kansas developed the first prototype of a GIS automated LESA model using Douglas County, Kansas as the study area. The reasons for this are obvious. First, it was handy, second there was a LESA model developed for the area and third, the vast majority of the data needed to implement the model were already available (at least in analog form).

For the current laboratory exercise you have a number of themes from an original Lee Williams (1985) study of a portion of Douglas County, Kansas. They have been converted from their original 'pMAP' format to ArcView Spatial Analyst format. The pMAP format corresponds to a raster GIS of the same name that is relatively similar to Tomlin's original Map Analysis Package. I believe it is still available. The themes that were not included in the database you are about to use are, by and large, the final results of the original analysis. The LESA model for Douglas County is described in detail in Williams' article, and you may want to familiarize yourself with it before you begin. Your task is to develop a final LESA model based on the themes provided for you. No detailed instructions for how to use Spatial Analyst will be included in this laboratory. You will now be expected to perform on your own.

Learning Objectives:

In this exercise you will gain further experience with Spatial Analyst. In addition you will gain experience with model formulation and model implementation. To assist your modeling efforts you will be required to flowchart both the formulation and the implementation. Your instructor may have specific requirements for flowcharting. You can flowchart by using pencil and paper (with or without a template), or you can use software for flowcharting. One piece of software called Chartist has proven useful to some students and is available for a free trial period. The URL for this package is http://www.Novagraph.com/. Currently ESRI is beta testing a landscape modeling interface for Spatial Analyst that will allow models to be flowcharted and then to be implemented from the flowchart itself.

When you are finished with this exercise you will be able to

- Develop a model without specific, step-by-step instructions.

- Develop a GIS model formulation.

- Flowchart the model formulation.

- Implement the completed cartographic model.

- Flowchart the model implementation.

Linkage to your text:

Your Fundamentals of Geographic Information Systems textbook covers cartographic modeling in detail in 13. You might want to refer back to it to get a better handle on the particulars of cartographic modeling and flowcharting. In addition, the notes for this course should be referred to so you begin to understand what this laboratory exercise is all about.

Technical Objectives:

1. To **<u>formulate</u>** a cartographic model flowchart indicating which themes are **needed**, rather than which are actually available for LESA modeling in Douglas County, Kansas. These should be based on the information provided in Williams' (1985) article.

2. To produce, from that flowchart, which themes are missing (if any). From this you are required to determine whether some modified themes might be used as surrogates for any missing themes. In addition, you are to determine which themes **must** be obtained through additional input. You might even consider explaining what sources you might use for these themes.

3. Based on the data available to you build as much of the LESA model as you can. Upon finishing, you will be required to explain how you performed the analysis (by giving a final flowchart of your implementation). The final flowchart may be different from the original one you developed to formulate the model.

Materials Provided:

1 Douglas County GIS Database in ArcView format (Each Grid Cell is 2.5 acres or 100 meters by 100 meters).

1 General description of the Douglas County LESA model based on the Williams (1985) article for guidance and background (Appendix A).

Products:

You are to provide the following information products from your analysis two (2) weeks from the date of the assignment.

1. One LESA model **formulation** flowchart. Remember that the flowchart moves in the opposite direction from the formulation itself.

2. One LESA model **implementation** flowchart. Remember here that this will not be the same as #1 because of the missing themes, or the surrogates you might have employed.

3. A 3 1/2" DSHD floppy disk with all of **your** themes (including intermediate themes) used to create your LESA model.

4. A detailed instruction set, similar to the ones you have been given in laboratory exercises 2 – 13, that could be used by anyone to reproduce what you created.

5. A 10 - 15 page write-up indicating what you did, how you dealt with the missing themes, how confident you are about your model, and describing the model implementation as if you are providing this to a client.

Exercise 14, Appendix A

LESA was an additive, linear planning model that creates a value between 0 and 300 points, with the higher values indicating that great effort should be put forth to preserve this land for agriculture. LESA came in two parts: Land Evaluation (LE) with a maximum value of 100 and Site Assessment (SA) with a maximum value of 300.

Land evaluation was a ranking of soils based on regional requirements for typical cropland types within the region. The soils were ranked, based on data from the National Coorperative Soil Survey and the best soils were assigned a value of 100, while soils of lesser value were assigned lower values. In this way, the better soils with higher ranks would be less likely to be developed for non-agricultural uses. To save you time, and because you don't have access to the Douglas County Soil Survey the following table shows the LE values for Douglas County. Use the Reclassify function under the Analysis pulldown menu to reclassify the soil series values to LE values.

Soil Series	Value	Label Range
Lake, Quarry, etc.	5	0 to 5
Wabash SC	40	26 to 40
Woodson SL 1-3%	85	71 to 85
Woodson SL 0-1%	70	56 to 70
Martin SCL 3-7%	55	41 to 55
Martin SCL 1-3%	85	71 to 85
Martin Soils 3-7%	40	26 to 40
Morrill CL 3-7%	100	86 to 100
Kennebec SL	100	86 to 100
Oska SCL 3-6%	70	56 to 70
Pawnee CL 3-7%	40	26 to 40
Pawnee CL 1-3%	85	71 to 85
Pawnee CL Eroded	25	16 to 25
Reading SL	100	86 to 100
Vinland Complex SL 3-7%	25	16 to 25
Vinland Martin Complex	15	6 to 15
Vinland Complex Eroded	15	6 to 15
Sogn-Vinland Complex	15	6 to 15
Stony Steep Land	5	0 to 5
Sibleyville L 3-7%	40	26 to 40
Gravelly Land	15	6 to 15
Gymer SL 1-3 & 3-8%	70	56 to 70
Wabash SCL	85	71 to 85

Site Assessment quantified other non-soils factors that contribute to the quality of land for agricultural use. Each selected factor was ranked based on both its importance to the model, and the degree to which it complied with that particular factor. For example one SA factor might be land in agriculture within 1 mile. This might have a relative weight of 10 (the highest relative **weight** that can be assigned). However, the percent of area in agriculture within 1 mile may have a number of values such as 0 to 30%, 31 to 70%, and 71 to 100%. Each of these may be ranked differently with, for example the 0 to 30% being assigned a low value of perhaps 3, 31 to 70% assigned a medium value of perhaps 7, and the 71 to 100% value assigned a value of 10. You might call this the **condition** of each factor.

Below are the SA factors, their weights and their compliance values. You will notice that not all the compliance values have been specified. You can decide what these values are. This gives you some control over what the final LESA maps will look like.

(1) Percent area in agriculture within 1.5 miles (Weight 8)

Value	Condition
10	95% of area in agriculture
---	50% of area in agriculture
0	10% of area in agriculture

(2) Land in agriculture adjacent to site (Weight 10)

Value	Condition
10	All sides of site in agriculture
---	One side of site adjacent to non-agricultural land
---	Two sides of site adjacent to non-agricultural land
---	Three sides of site adjacent to non-agricultural land
0	The site surrounded by non-agricultural land

(3) Size of farm (Weight 7)

Value	Condition
10	120 acres or more
---	80 to 120 acres
---	40 to 80 acres
---	20 to 40 acres
---	10 to 20 acres
0	Less than 10 acres

(4) Average parcel size within 1 mile of site (Weight 9)

Value	Condition
10	120 acres or more
---	80 to 120 acres
---	40 to 80 acres
---	20 to 40 acres
---	10 to 20 acres
0	Less than 10 acres

(5) Agrivestment in real property improvements within 2 miles (Weight 10)

Value	Condition
10	High level of investment in farm facilities (long-term)
---	Moderate level of investment
0	Diminishing level of investment

(6) Percent of land zoned agriculture within 1.5 miles (Weight 8)

Value	Condition
10	90 or more
---	75 to 89
---	50 to 74
---	25 to 49
0	Less than 25

(7) Zoning of the site and adjacent to it (Weight 6)

Value	Condition
10	Site and all surrounding sides zoned for agricultural uses
---	Site zoned agricultural and one side zoned low density residential
---	Site zoned agricultural and two sides zoned for residential, commercial, or industrial
0	Site surrounded by residential, commercial, or industrial zoning

138

(8) Availability of land zoned for proposed use (Weight 6)

Value	Condition
10	Undeveloped land zoned for proposed use is beyond the primary and suburban growth areas of the incorporated cities
0	No zoned land available for proposed use (this value can only be assigned when a parcel is within the primary or suburban growth areas)

(9) Availability of non-farmland or less productive land as alternative site within area (Weight 6)

Value	Condition
10	Large amount available
---	Moderate amount available
0	None available

(10) Need for additional urban land (Weight 8)

Value	Condition
10	Vacant, buildable land within city limits, capable of accommodating proposed use
0	Little or no vacant land remaining within the city limits to accommodate the proposed use

(11) Compatibility of proposed use with the surrounding area (Weight 7)

Value	Condition
10	Not compatible – high intensity uses
---	Somewhat compatible but not totally
0	Compatible

(12) Unique topographic, historic, or ground cover features or unique scenic qualities (Weight 3)

Value	Condition
10	All of the site
---	Part of the site
0	None of the site

(13) Adjacent to land with unique topographic, historic, or ground cover features or unique scenic qualities (Weight 2)

Value	Condition
10	On all sides of the site
---	Three sides of the site
---	Two sides of the site
---	One side of the site
0	None of the site is adjacent to these unique features

(14) Site subject to flooding or in a drainage course (Weight 8)

Value	Condition
10	All of the site
---	50 percent of the site
0	None of the site

(15) Suitability of soils for on-site waste disposal (Weight 5)

Value	Condition
10	All of the site
---	50 percent of the site
0	None of the site

(16) Compatibility with an adopted comprehensive plan (Weight 5)

Value	Condition
10	Soil limitation that restricts the use of septic system
---	Limitations to the soil can be overcome by special management
0	Little or no limitations

(17) Within a designated growth area (Weight 5)

Value	Condition
10	Rural area
---	Clinton reservoir sanitation zone
---	Suburban growth area
0	Primary growth area

(18)   Distance from city limits (Weight 6)

Value	Condition
10	More than 2 miles
---	2 miles or less
---	1.5 miles or less
---	1 mile or less
---	0.5 miles or less
0	Adjacent

(19)   Distance from transportation (Weight 5)

Value	Condition
10	Limited transportation access dominated by rural township roads
---	Access to improved county roads or highway within suburban growth areas
---	Access to improved county roads or highway within primary growth areas
0	Access to full range of transportation services

(20)   Distance from central water system (Weight 4)

Value	Condition
10	No water within 1 mile
---	Water within 2000 feet
0	Water at the site

(21)   Distance from sewage lines (Weight 4)

Value	Condition
10	No sewer line within 1.5 miles
---	Sewer line within 1 mile
---	Sewer line within 0.5 miles
0	Sewer line adjacent to site